Mastering Veeam Backup & Replication 10

Protect your virtual environment and implement cloud backup using Veeam technology

Chris Childerhose

BIRMINGHAM—MUMBAI

Mastering Veeam Backup & Replication 10

Group Product Manager: Wilson D'suoza
Associate Publishing Product Manager: Sankalp Khattri
Senior Editor: Shazeen Iqbal
Content Development Editor: Ronn Kurien
Technical Editor: Yoginee Marathe
Copy Editor: Safis Editing
Project Coordinator: Ajesh Devavaram
Proofreader: Safis Editing
Indexer: Pratik Shirodkar
Production Designer: Joshua Misquitta

First published: March 2021
Production reference: 1160221

Published by Packt Publishing Ltd.
Livery Place
35 Livery Street
Birmingham
B3 2PB, UK.

ISBN 978-1-83898-044-3

www.packt.com

Contributors

About the author

Chris Childerhose is an information technology professional with over 25 years of experience in network/systems architecture, network and systems administration, and technical support. He is a Veeam Vanguard, a Veeam Certified Architect, and a Veeam Certified Engineer. He also has the following certifications: vExpert, VCAP-DCA, VCP-DCV, and MCITP. He currently works for ThinkOn as the lead infrastructure architect, where he designs the infrastructure for all client services offered. Chris is also an avid blogger on all things virtual, focusing on Veeam and VMware.

Writing a book is harder than I thought and more rewarding than I could have ever imagined. None of this would have been possible without my wife, Julie. She stood by me during every struggle and all my successes, including many nights writing. She always pushes me in my career endeavors and I will be forever thankful for this.

I'm eternally grateful to Rick Vanover of Veeam, who took the time to do a technical review of all my chapters when he didn't have to. He was great at suggesting changes or edits and even contributed to a couple of chapters with his perspective. I truly have no words to thank him for this amazing selflessness to help a fellow Veeamer.

To all my colleagues and fellow Vanguards who spoke words of encouragement and praise about me writing a book, it is truly humbling to have a great community in the Veeam Vanguards as well as colleagues that support your work.

About the reviewer

Robert Verdam is a very experienced technical consultant with extensive experience in designing, installing, configuring, supporting, and troubleshooting a wide range of cloud and on-premises solutions. Robert is also a skilled developer and he breathes automation. His developing skills mostly focus on PowerShell and Python development.

His broad certifications include VMware VCIX-DCV (VMware Certified Implementation Expert for Datacenter Virtualization) and Cisco CCNP (Cisco Certified Network Professional). He currently holds the VMCE 2020 (Veeam Certified Engineer) certification and is one of the few Dutch Veeam certified architects (VMCA), and as such, he is also very knowledgeable on the Veeam product portfolio and all related products.

Table of Contents

Section 2: Storage – NAS Backup, Linux, SOBR, and OBS

3

NAS Backup

4

Scale-Out Repository and Object Storage – New Copy Policy

5

Windows and Linux – Proxies and Repositories

6

Object Storage – Immutability

Section 3: DataLabs, Cloud Backup, and Veeam ONE

7

Veeam DataLabs

8

Cloud Backup and Recovery Using Veeam Cloud Connect Provider and the Insider Protection Feature

9

Instant VM Recovery

10

Veeam ONE

Other Books You May Enjoy

Index

Preface

Veeam is one of the leading modern data protection solutions, and learning this technology can help you protect your virtual environments effectively. This book guides you through implementing modern data protection solutions for your cloud and virtual infrastructure with Veeam. You will even gain in-depth knowledge of advanced-level concepts such as DataLabs, Cloud Backup and Recovery, Instant VM Recovery, and Veeam ONE.

Who this book is for

Readers of this book will be VMware administrators or backup administrators with some existing knowledge of Veeam and the topics covered in this book. They will know some virtualization and backup concepts to understand what is discussed in each chapter. Veeam is one of the leading modern data protection solutions, and learning this technology will help users protect their environments. Most readers will want to implement many of the topics discussed and investigate Veeam ONE for monitoring/reporting their infrastructure.

What this book covers

Chapter 1, Installation – Best Practices and Optimizations, will cover the installation, best practices, and optimizations for Veeam Backup & Replication. You will learn how to set up things such as the backup server, proxies, repositories, and more. There will be a reference to the Veeam Best Practice site as well. Once best practices are covered, we will dive into further optimizations with repository servers, proxy servers, and more.

Chapter 2, The 3-2-1 Rule – Keeping Data Safe, will discuss the all-important 3-2-1 rule of backups: three copies of data, two on different media, and one offsite. We will discuss the importance of keeping your data safe, and using this method is one of the best ways to do this. We will discuss some of the different media types that we can use, including tape for air-gapped protection.

Chapter 3, *NAS Backup*, will cover the powerful new NAS backup feature of Veeam 10. We will look at the flexibility of the different protocols supported, **CBT** (short for **Changed Block Tracking**), and snapshot-friendly backups. We will dive into how to configure NAS shares on the Veeam server, create a backup job, and even go through the restore process.

Chapter 4, *Scale-Out Repositories and Object Storage – New Copy Policy*, will dive into scale-out repositories and best practices for creating them. We will also discuss different filesystems, such as ReFS and XFS. Capacity Tier will be addressed as a way to manage a **SOBR** (short for **Scale-Out Backup Repository**) capacity by offloading data. Included in this will be the new copy policy in version 10 that tells Veeam to offload to Capacity Tier as soon as the backup files are on the SOBR.

Chapter 5, *Windows and Linux – Proxies and Repositories*, will dive into the new Linux repository options using Reflink and Fast Clone and the Windows repositories. We will discuss recommended Linux versions, set up Reflink, and then configure the XFS repository in Veeam. We will show how to enable Fast Clone and the recommended best settings when creating a job for this storage to take advantage of the Fast Clone technology. We will dive into the world of proxy servers – the workhorses for Veeam Backup. We will discuss the pros and cons of using Windows versus Linux proxies and where each of them plays a role. We will then discuss best practices for setting them up and configuration, as well as the different modes that each of them can use during the backup process, including snapshot integration.

Chapter 6, *Object Storage – Immutability*, will discuss the rise of object storage and the role that it plays with Veeam. We will discuss how to configure and use it within Veeam, including the Immutability options. We will also discuss some vendor offerings, including WORM options for object storage. We will even touch on how this can also be considered air-gapped protection for your backups.

Chapter 7, *Veeam DataLabs*, will dive into one of the great features of Veeam: DataLabs. We will discuss how to set it up, including the three main requirements: a virtual lab, application groups, and SureBackup jobs. Each of these is addressed in detail, including settings and best practices. We will also discuss how DataLabs can help dev/QA departments test new applications or patches and test new software. We will even touch briefly on storage snapshot integration.

Chapter 8, *Cloud Backup and Recovery Using Veeam Cloud Connect Provider and the Insider Protection Feature*, will discuss another way for users to protect their data by using a Veeam Cloud Connect provider and the Insider Protection feature. We will discuss how this can protect data from ransomware and how it gets configured. We will look at Veeam Backup & Recovery options for cloud providers and why they should look into this as an option. We will explain how this also protects users from an insider attack on the Veeam infrastructure and how it can save backup files.

Chapter 9, Instant VM Recovery, will discuss what Instant VM Recovery is and how it works. We will dive into why a user would want to use it for recovery purposes directly to VMware in vCenter or ESXi hosts. We will explain the benefits of using Instant VM Recovery for testing purposes and how you can restore VMs for the dev or QA testing of applications, patches, and so on.

Chapter 10, Veeam ONE, will cover Veeam's monitoring and reporting utility: Veeam ONE. We will discuss how Veeam ONE can help monitor your environment from backup servers, VMware, and vCloud Director. We will also discuss the reporting aspect and schedule reports to help with your daily administration tasks. We will also see how Veeam ONE can benefit any organization using Veeam.

To get the most out of this book

You should have at least 6 months of hands-on knowledge of using Windows/Linux servers as well as some experience of virtualization with VMware. It would be best if you were comfortable with setting up servers and configuring them with storage. You should also have some backup knowledge and have already used Veeam, even for basic tasks, since many topics in the book look at the more advanced features of Veeam Backup & Replication.

Software/hardware covered in the book	OS requirements
Windows 2016/2019	Windows
Veeam Availability Suite	Windows
Ubuntu Linux 20.04	Linux

You also need Windows Server set up to install Veeam Backup & Replication and Veeam ONE. You can download the Veeam Availability Suite ISO file and trial license from `http://www.veeam.com`.

If you are using the digital version of this book, we advise you to type the code yourself. Doing so will help you avoid any potential errors related to the copying and pasting of code.

Download the color images

We also provide a PDF file that has color images of the screenshots/diagrams used in this book. You can download it here: `http://www.packtpub.com/sites/default/files/downloads/9781838980443_ColorImages.pdf`.

Conventions used

There are a number of text conventions used throughout this book.

`Code in text`: Indicates code words in text, database table names, folder names, filenames, file extensions, pathnames, dummy URLs, user input, and Twitter handles. Here is an example: "The following is the datastore mounted to the host, `home-esxi02.home.lab`, for the Instant VM Recovery process."

A block of code is set as follows:

```
html, body, #map {
  height: 100%;
  margin: 0;
  padding: 0
}
```

When we wish to draw your attention to a particular part of a code block, the relevant lines or items are set in bold:

```
[default]
exten => s,1,Dial(Zap/1|30)
exten => s,2,Voicemail(u100)
exten => s,102,Voicemail(b100)
exten => i,1,Voicemail(s0)
```

Any command-line input or output is written as follows:

```
$ mkdir css
$ cd css
```

Bold: Indicates a new term, an important word, or words that you see onscreen. For example, words in menus or dialog boxes appear in the text like this. Here is an example: "The first thing to create is the virtual lab, so select the **ADD VIRTUAL LAB** option toward the right-hand side of the screen."

> **Tips or important notes**
> Appear like this.

Get in touch

Feedback from our readers is always welcome.

General feedback: If you have questions about any aspect of this book, mention the book title in the subject of your message and email us at customercare@packtpub.com.

Errata: Although we have taken every care to ensure the accuracy of our content, mistakes do happen. If you have found a mistake in this book, we would be grateful if you would report this to us. Please visit www.packtpub.com/support/errata, selecting your book, clicking on the Errata Submission Form link, and entering the details.

Piracy: If you come across any illegal copies of our works in any form on the Internet, we would be grateful if you would provide us with the location address or website name. Please contact us at copyright@packt.com with a link to the material.

If you are interested in becoming an author: If there is a topic that you have expertise in and you are interested in either writing or contributing to a book, please visit authors.packtpub.com.

Reviews

Please leave a review. Once you have read and used this book, why not leave a review on the site that you purchased it from? Potential readers can then see and use your unbiased opinion to make purchase decisions, we at Packt can understand what you think about our products, and our authors can see your feedback on their book. Thank you!

For more information about Packt, please visit packt.com.

Section 1: Installation – Best Practices and Optimizations

The objective of this section is to teach the reader the best practices and optimizations for installing Veeam. This section will also include the 3-2-1 backup rule for safeguarding data. You will be able to apply the best practices and optimizations to your installation of Veeam.

This section contains the following chapters:

- *Chapter 1, Installation – Best Practices and Optimizations*
- *Chapter 2, The 3-2-1 Rule – Keeping Data Safe*

1
Installation – Best Practices and Optimization

Veeam Backup & Replication v10 is part of **Veeam Availability Suite**, which is ready for the modern data center and allows you to back up all of your workloads, including **Cloud**, **Virtual**, and **Physical**. It is simple, yet flexible, when it comes to meeting your most challenging business requirements. In this chapter, we will discuss the installation of the software, what components make up Veeam Backup & Replication v10, and some best practices and optimizations. There will be practical examples throughout the chapter of how to optimize specific elements that make up the **Veeam** environment. We will also touch on some of the websites, including the *Best Practices Guide for Veeam*, among others, to give you the resources to help set up Veeam in your environment. As they say with Veeam – *It Just Works*.

In this chapter, we're going to cover the following main topics:

- Understanding the components of Veeam Backup & Replication
- Understanding the best practices for Veeam installation and setup
- Configuring and optimizing proxy servers
- Setting up repository servers for success
- Understanding the scale-out repository

Technical requirements

To ensure a successful installation, you will require the following:

- A *Windows 2016/2019* server with the required disk space to install the application (*2012 R2* is also currently supported). *Windows 10* and other modern *Windows* desktop operating systems are also supported.
- The latest **ISO file** from www.veeam.com, which requires registration on the site and allows you to obtain a trial license. As of the time of writing, version 10.0.1.4854 is the current release.

Understanding the components of Veeam Backup & Replication

The Veeam Backup & Replication software has several components that together make up the complete architecture that is needed to protect your environment.

Veeam Backup & Replication has the following components:

- **Backup Server**: Installed on either a physical or virtual server, this is the core component of Veeam Backup & Replication that controls and coordinates backups, replication, recovery verification, and restore tasks. It manages job scheduling as well as resource allocation. It also contains global configuration settings for the environment.

- **Proxy Server**: These are the workhorses of the environment as they offload tasks from the backup server and are the data movers between the backup server and repositories. It is the proxy servers that you can scale to add processing tasks to the environment.

- **Repository Server**: This is the backup target where all backup files (VBM – metadata, VBK – full backup file, and VIB – incremental) get written. The repository servers can be Windows- or Linux-based and have different filesystems, such as NTFS, ReFS, and XFS.

- **Enterprise Manager Server**: This server is an optional component and gets installed when you want to manage multiple backup servers in a single pane of glass. You can see backup jobs within your environment from multiple backup servers. Enterprise Manager also allows you to search all Microsoft Windows guest OS files within all current and archived backups with one-click restore. This optional component uses a separate SQL server and backup catalog service for indexing the guest operating systems.

- **Built-In WAN Acceleration**: This is also an optional component that allows for better movement of data between sites. It helps minimize data transfer by comparing the data blocks before the transfer, so only new blocks are sent across the WAN. It also accelerates backups by up to 50x between sites.

- **File Explorers**: These are built-in applications used during restore and look very similar to Windows Explorer. They allow you to browse the restore point to select specific files or data to be restored. There are explorers for Active Directory, Microsoft SQL, Oracle, Microsoft Exchange, Microsoft SharePoint, and Microsoft OneDrive for Business.

You can reference the Veeam Backup & Replication File Explorers at the following website:

```
https://helpcenter.veeam.com/docs/backup/explorers/explorers_
introduction.html?ver=100
```

The following diagram will show all the aforementioned components:

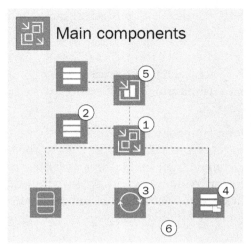

Figure 1.1 – Veeam Backup & Replication components

In a simple setup, where all components are installed on one server but can scale as needed, you will require at least the backup server, proxy server, and repository server.

When you have multiple offices, you may wish to deploy Veeam Backup & Replication in a more advanced setup, as illustrated here:

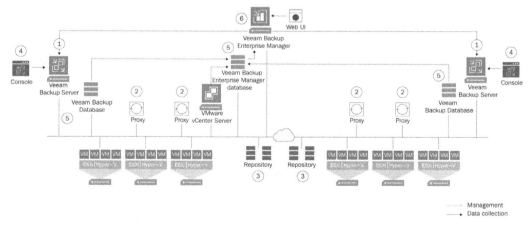

Figure 1.2 – Advanced or distributed architecture installation

This deployment depicts the advanced or distributed setup of the application across more than one office where Enterprise Manager would see both backup servers.

Now that you have a better understanding of the components that make up Veeam Backup & Replication, we will now get into the installation, as well as best practices and optimizations, in the next section.

Understanding the best practices and optimizations for Veeam installation

The installation of **Veeam Backup & Replication v10** is a straightforward process, and this section will go through the operation of the install as well as touch on best practices and optimizations for your environment. Setting up Veeam, if not done right, can lead to components not working correctly and poor performance, among other things. However, if you set up **Veeam** correctly, it will protect your data and environment with little configuration.

Installation of Veeam Backup & Replication v10

Before installing **Veeam**, we need to ensure that you have a server deployed, either *Windows 2016 or 2019*, with enough disk space for the installation. The disk layout should be similar to the following:

- **OS drive**: This is where your operating system resides and should be used only for this purpose.

- **Application drive**: This will be your application installation drive for Veeam and all its components.

- **Catalog drive**: Veeam uses a catalog that can generate around 10 GB of data per 100 VMs backed up with file indexes. If this were to be a significant storage requirement for your deployment, it may be advisable to allocate to a separate drive.

Once your server is ready, and you have downloaded the **ISO file** and mounted it, complete the following steps for installation:

1. Run the `setup.exe` file on the mounted ISO drive:

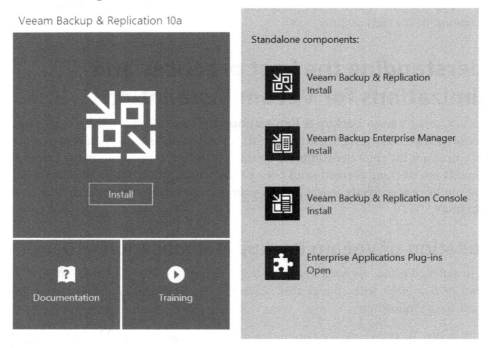

Figure 1.3 – Main installation screen

2. Click either on the **Install** button under the **Veeam Backup & Replication 10a** section on the left or the **Install** link on the right side under **Standalone components**. At this point, you will see the **License Agreement** window, so you need to select the two checkboxes to place a checkmark and then click **Next** to continue.

3. You will now need to provide a valid license file, be it a purchase or a trial. If you do not have it at this part of the installation, you can click **Next** to continue, and Veeam will operate in the *Community (Free) Edition*. When you obtain the license file, you can install that within the application:

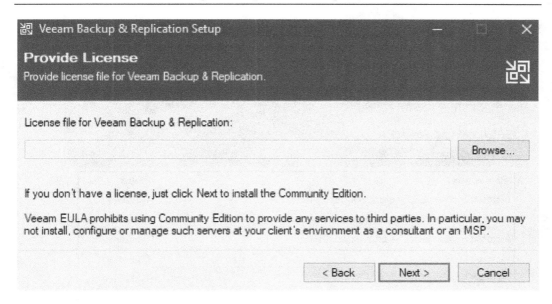

Figure 1.4 – License dialogue window

4. The next screen is where you choose which components you want to install and which directory. Veeam recommends that all of them are selected:

    – **Veeam Backup & Replication**: The main application.

    – **Veeam Backup Catalog**: Used when you turn on **Guest OS Indexing** within your jobs. This option takes all the *Guest* files and stores them in a catalog where you are then able to use advanced searching across all restore points and conduct 1-click file restores from the Enterprise Manager console.

    – **Veeam Backup & Replication Console**: This is where you go to view, create, and edit jobs, and manage the environment.

5. After clicking **Next**, the installer will then do a system check for any pre-requisites required, and if something is missing, you will be prompted and have the option to install the missing components:

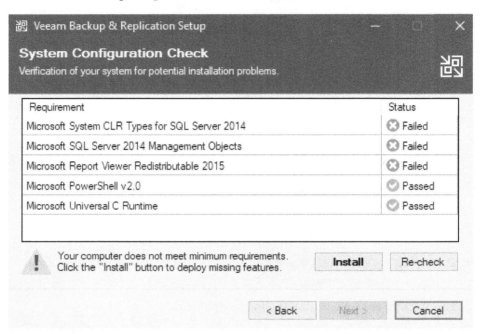

Figure 1.5 – System Configuration Check – missing components

6. Click the **Install** button to have the missing components installed.

7. Once all the components have passed, you can click **Next** to move to the following screen. Unlike in previous versions of Veeam Backup & Replication, the next screen does not give you the option to input a user account to run the services. Instead, with version 10 of Veeam, you need to select the checkbox next to **Let me specify different settings** and then click **Next**.

8. You will now have the opportunity to enter a user account for the Veeam services, better known as a *Service Account*. There are some recommended settings for this service account:

    – You must have *Local Administrator* rights on the Veeam server.

    – If you are using a separate SQL Server and not the Express edition that comes with the install, you require permissions to create the database.

    – You will need full NTFS permissions to the folder containing the catalog.

    For all the detailed permissions, please visit `https://helpcenter.veeam.com/docs/backup/vsphere/required_permissions.html?ver=100`:

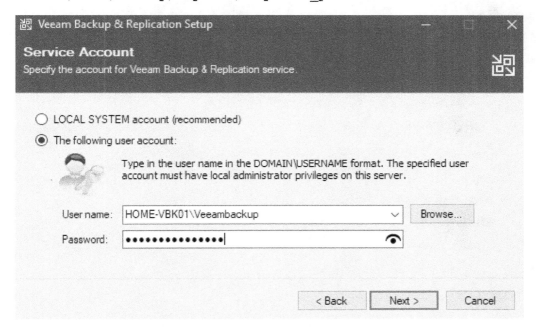

Figure 1.6 – User account for services

For this setup, I am using an account that I created on my lab server. In contrast, in a production scenario, you would already have a service account set up in Active Directory to enter at this step:

1. The next screen lets you select the type of SQL install you will be using, and in a lab scenario, using SQL Express is good enough. If you are in an enterprise environment, the recommended best practice is to use an external SQL Server for best performance. Also, note that you can use **Windows authentication** or **SQL Server authentication**:

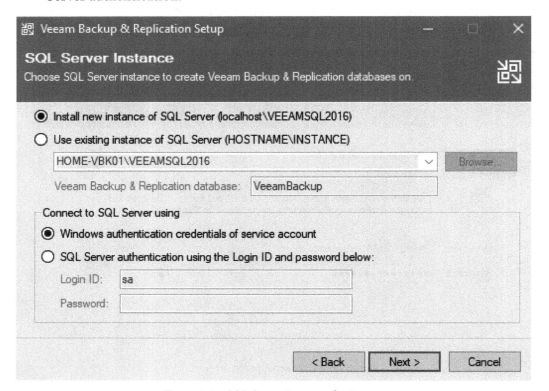

Figure 1.7 – SQL Server instance for Veeam

2. After selecting the appropriate options, click **Next** once again.

3. The next window is the TCP/IP port configuration. Should you want to use different ports, you can adjust them here, but the default ports should suffice. You then click **Next** to move to the **Data Locations** screen:

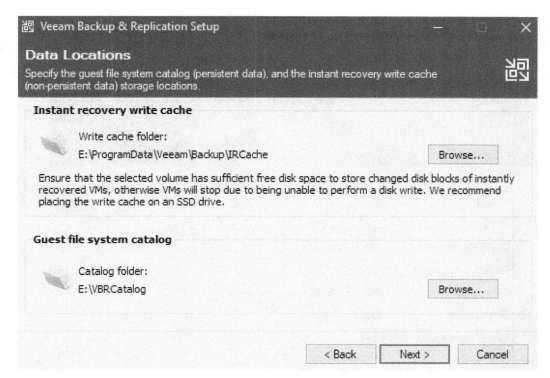

Figure 1.8 – Data Locations – directory selection

4. Here, you indicate the application drive for the **Instant recovery write cache**, which mounts restore points during recovery and the dedicated drive for **Catalog folder** for guest OS indexing.

   The installer is now ready to first complete installation of the local SQL Express instance and then the application. Veeam will also set your user account that you selected to initiate all the services:

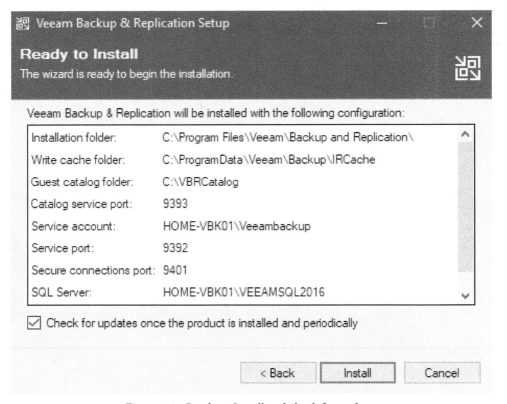

Figure 1.9 –Ready to Install and check for updates

5. After reviewing the setup, click **Install** to proceed with the installation and start setting up the components that work together with the backup server.

We will now proceed to configuring the required settings for Veeam to work with VMware:

- **Repository Server**: The server used to store the backup files.

- **Proxy Servers**: The servers that perform all the backup tasks.

- **VMware vCenter Credentials**: Used to connect and see your clusters, hosts, vApps, and Virtual Machines (VMs). vCenter Server is not required as standalone ESXi hosts are also supported if licensed in VMware:

1. When you first launch the **Veeam Backup & Replication** console, you are taken directly to the **Inventory** tab, and **Virtual Infrastructure** will be highlighted:

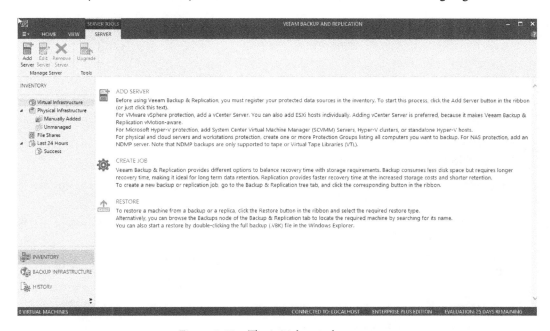

Figure 1.10 – The initial console screen

2. This screen is where we will begin adding the virtual center to enable the backup of your virtual machines. Click on the **ADD SERVER** option to start the process. You will then get prompted to select what kind of server to add. Choose **VMware vSphere** and then either **vSphere** or **vCloud Director**:

Figure 1.11 – vSphere or vCloud Director selection

You would typically select **vSphere**; however, if you have vCloud Director in your environment, you may also want to choose this option. When you choose **vSphere**, you will get prompted for two things to complete the connection:

– The DNS or IP address of your vCenter server.

– Credentials; these can either be a vsphere.local user or a domain account set up for access.

> **Important note**
> You do have the option of selecting Microsoft Hyper-V and Nutanix AHV, but this book is focused mainly on VMware vSphere.

3. Enter in the required credentials and then click **Next**, followed by **Apply**, to complete the VMware vSphere setup. You will now see your vCenter server listed under the **Virtual Infrastructure** section of the console and will be able to browse the hosts and virtual machines.

We now move on to the next piece required for the infrastructure, which is the *proxy server*. By default, the **Veeam Backup & Replication** server is to be your *VMware Backup Proxy* and *File Backup Proxy*. Due to the limitations of my lab, I am going to use this server as an example, but in the real world, you would add multiple proxy servers to your environment for better performance and according to best practice. Also, based on best practices, you would typically disable the Veeam Backup & Replication server from being a proxy server to allow the other proxy servers to handle the workload.

The next component you will require is a *Repository Server*, the location where Veeam Backup & Replication will store the backup files. By default, Veeam Backup & Replication creates a *Default Backup Repository*, typically on the most significant sized drive attached to your backup server. This location will be where the *Configuration Backups* usually get backed up. There are multiple options for adding a repository:

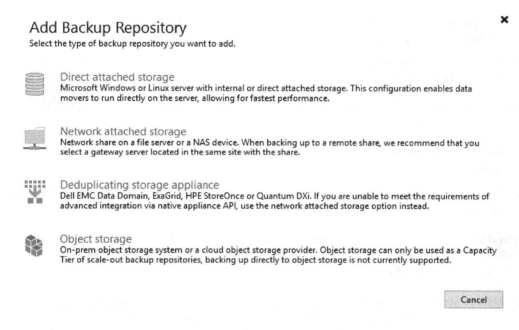

Figure 1.12 – Backup Repository selection

The first three selections would be for block storage, and the last one is object storage, which only works when creating a scale-out backup repository as the capacity tier for offloading data.

For further information on Veeam Backup & Replication best practices and the documentation, please visit the following websites:

- Veeam best practices website: `https://bp.veeam.com/vbr/`

- Veeam documentation website: `https://helpcenter.veeam.com/docs/backup/vsphere/overview.html?ver=100`

You have now completed the installation and basic configuration required for **Veeam Backup & Replication**. We will now look at how to optimize the proxy and repository servers in the next sections.

Configuring and optimizing proxy servers

Proxy servers are the workhorses of the Veeam Backup & Replication v10 application, and they do all the heavy lifting or processing of tasks for backup and restore jobs. When you set up Veeam, you need to ensure that the proxy servers get configured as per best practices:

- `https://bp.veeam.com/vbr/VBP/2_Design_Structures/D_Veeam_Components/D_backup_proxies/vmware_proxies.html`

- `https://helpcenter.veeam.com/docs/backup/vsphere/backup_proxy.html?ver=100`

When you decide to deploy a proxy server, Veeam Backup & Replication will install two components on the server:

- **Veeam Installer Service**: This is used to check the server and upgrade software as required.

- **Veeam Data Mover**: This is the processing engine for the proxy server and performs all the required tasks.

Veeam Backup & Replication proxy servers use what we call a *Transport Mode* to retrieve data during backup. Three standard modes are available, and they are listed in order, starting with the most efficient method:

- **Direct Storage access**: The proxy is placed in the same network as your storage arrays and can retrieve data directly from there. This method allows for two transport modes – Direct SAN access and Direct NFS access. The backup load is offloaded from the hypervisor to process the workloads.

- **Virtual Appliance**: This mode mounts the VMDK files to the proxy server for what we typically call **Hot-Add Mode** to back up the server data.

- **Network**: This mode is the least efficient, and is used when the **Failover to network mode if primary mode fails, or is unavailable** option is selected. It moves the data through your network stack, and it is recommended not to use 1 GB, but rather 10 GB or higher.

In addition to these standard transport modes provided natively for VMware environments, Veeam provides two other transport modes: Backup from Storage Snapshots, and Direct NFS. These provide storage-specific transport options for NFS systems and storage systems that integrate with Veeam.

Refer to the integration with storage systems guide: `https://helpcenter.veeam.com/docs/backup/vsphere/storage_integration.html?ver=100`.

Along with the transport modes, there are specific tasks that the proxy server performs:

- Retrieving the VM data from storage
- Compressing
- Deduplicating
- Encrypting
- Sending the data to the backup repository server (backup job) or another backup proxy server (replication job)

Veeam proxy servers leverage what is known as **VADP (VMware vStorage APIs for Data Protection)** when using all transport modes other than Backup from Storage Snapshots and Direct NFS.

The following are things you should consider in relation to your proxy servers:

- **Operating System**: Most software vendors will always recommend the latest and greatest, so if you are going to choose Windows, then 2019, or if you are going to choose Linux, then the newest flavor you have picked (*Example – Ubuntu 20.04.1 LTS*). Note that as regards Linux VMware backup proxies, only *HotAdd* mode is supported in Veeam Backup & Replication v10.
- **Proxy Placement**: Depending on the transport mode for the server, you will need to place it as close to the servers you want to back up, such as on a specific host in VMware, a physical server, or a blade enclosure. The closer to the source data, the better!
- **Proxy Sizing**: This can be tricky to determine and will be dependent on the server being physical or virtual. Veeam proxy servers complete what are called *Tasks*, which is the processing of one virtual disk for a VM or one physical disk for a server. Therefore, Veeam recommends one physical core or one vCPU as well as 2 GB of RAM per task.

Veeam has a formula used to calculate the required resources for a proxy server:

- D = source data in MB
- W = backup window in seconds
- T = throughput in MB/s, = *D/W*
- CR = change rate
- CF = cores required for full backup, = *T/100*
- CI = cores required for incremental backup, = *(T * CR)/25*

Based on these requirements, we can use a sample of data to perform the calculations:

- 500 virtual machines
- 200 TB of data
- 8-hour backup window
- 10% change rate

Using these numbers, we perform the following calculations:

$$D = 200\,TB * 1024 * 1024 = 209,715,200\,MB\;(\textbf{\textit{Data is converted to MB}})$$

$$W = 8\,hours * 3600\,seconds = 28,800\,seconds$$

$$T = 209,715,200 / 28,800 = 7,282\,MB/s$$

This formula determines the throughput required for the data that will be ingested by the backups.

We now use the numbers we calculated to determine the required number of cores needed to run both full backup and incremental backup to meet your defined SLA:

$$CF = T / 100\;\;CF\,(\textbf{\textit{Full}}) = 7,282 / 100 \sim 73\,cores$$

$$CI = (T * CR) / 25\;\;CI\,(\textbf{\textit{Incremental}}) = (7,282 * 10) / 25 \sim 29\,cores$$

This formula takes the throughput from the previous formula and then calculates the number of CPU cores required.

Based on our calculations and considering you require 2 GB of RAM for each task, you would need a virtual server with 73 vCPUs and 146 GB of RAM. This size may seem like a considerable server, but keep in mind that it uses the sample data. Your calculations will likely be much smaller or possibly more extensive, depending on your dataset.

Should you decide to use a physical server as a proxy, you should have a server with 2 – 10 core CPUs. In the case of our sample data, two physical servers are what you require. If you are using virtual servers for proxies, then the best practice is to configure them with a maximum of 8 vCPUs and add as many as required for your environment – in this case, we would need nine servers.

Should you want to size things based on incremental backups only, your requirements are less than half of those for a full backup – 29 vCPUs and 58 GB of RAM.

There are limitations for proxy servers that you need to be aware of when it comes to job processing and performance. As noted above, a proxy server performs *tasks*, which are assigned CPU resources. Concurrent task processing is dependent on the resources you have available in your infrastructure and the number of proxy servers you have deployed. As seen here, when adding a proxy server to Veeam Backup & Replication, there is the **Max concurrent tasks** option, which correlates to the number of CPUs that are assigned:

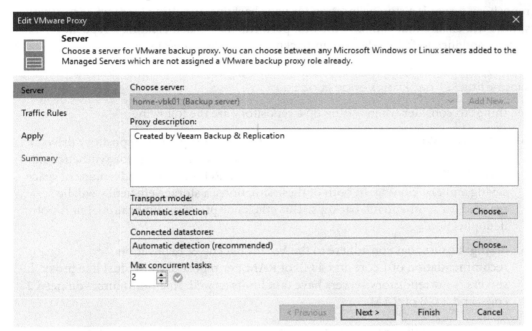

Figure 1.13 – Max concurrent tasks limitation for a proxy

The task limits can be viewed at the following link: `https://helpcenter.veeam.com/docs/backup/vsphere/limiting_tasks.html?ver=100`.

Important note

Job performance gets impacted based on the tasks of a proxy server. As an example, if you had a proxy server with 8 CPUs and you added 2 virtual machines for backup, one with 4 disks and the other with 6 disks, only 8 of 10 disks would get processed in parallel. The remaining 2 disks would have to wait on resources before backing up because tasks get assigned per virtual disk of a VM during the backup process.

You should now be able to size your proxy servers correctly regarding things such as CPUs and RAM and understand proxy placement and how it processes tasks. Proxy servers send data to repository servers, which is the focus of the next section.

Setting up repository servers for success

A repository server is a storage location for your backups, so setting them up correctly the first time will ensure that you have the best performance. When creating a repository, it is always a good idea to follow the Veeam Backup & Replication best practices:

```
https://bp.veeam.com/vbr/VBP/2_Design_Structures/D_Veeam_
Components/D_backup_repositories/
```

Some things to consider when setting up a repository are the following:

- **ReFS/XFS**: With Windows 2016/2019, ensure you format your repository drive(s) as ReFS with 64k block sizing to take advantage of space savings for synthetic fulls and GFS. For Linux, you need to set up XFS and Reflink to take advantage of space saving and fast cloning. In both of these situations, a storage efficiency will be realized for synthetic full backups. This efficiency prevents duplication, but is not deduplication.

- **Sizing**: Ensure that you adhere to the Veeam Backup & Replication recommendation of 1 core and 4 GB of RAM per repository task. Just like proxy servers, your repository servers have task limits as well. At a minimum, you need 2 cores and 8 GB of RAM.

When you calculate out the sizing requirements, you need to take into account your proxy servers and the amount of CPUs configured. You then need to use a 3:1 ratio for the core count on a repository server:

Example: If your proxy server was configured with 8 CPUs, you would then need to configure the repository server with 2 CPUs based on this rule of 3:1. To configure the RAM, you multiply the CPU count by four, ending up with 8 GB of RAM.

When you use the Windows ReFS filesystem as your repository, you need to consider the overhead required for the filesystem and be sure to add another 0.5 GB of RAM per terabyte of ReFS.

Setting up your task limits for a repository server is different to a proxy server due to the way tasks are consumed. The setting chosen will be handled differently:

- **Per-VM Backup Files**: When selected, this creates a backup chain per VM located in a job. Therefore, if the backup job has 10 virtual machines, then it will consume 10 repository tasks as well as 10 proxy tasks.

- **No Per-VM Selection**: The backup job consumes one repository task because all VM data gets written to the same backup file, and the proxy task will remain the same with one task per each virtual disk.

The task limits can be viewed at the following link: `https://helpcenter.veeam.com/docs/backup/vsphere/limiting_tasks.html?ver=100`.

When setting up a repository for the first time, you can set the task limit as follows:

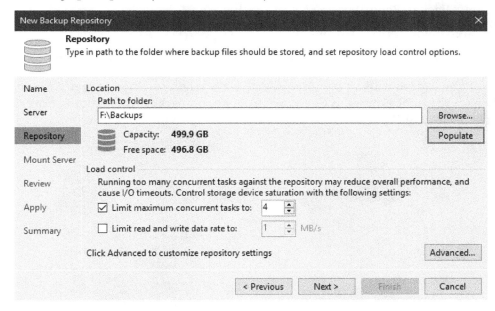

Figure 1.14 – Repository task limit

Important note

When you limit the number of tasks per repository, and you have jobs with many virtual machines requiring backup, this will be one of the bottlenecks in your environment. You also need to ensure that you do not set the limit too high as that could overwhelm your storage, causing performance degradation. Make sure to test all your components and resources available for your backup infrastructure.

You should, after this section, be able to choose which type of filesystem for your repository and also size it correctly based on CPU and RAM. We also discussed the per-VM versus no per-VM methods. Now we will use this knowledge to tie this into creating a scale-out repository.

Understanding the scale-out repository

So, what is a **Scale-Out Backup Repository**, or **SOBR**, you ask? A SOBR, uses multiple backup repositories called performance extents to create a sizeable horizontal scaling repository system. Veeam Backup & Replication can use multiple repositories of various types, such as the following:

- **Windows Backup Repositories**: NTFS or the recommended ReFS

- **Linux Backup Repositories**: XFS with Reflink

- **Shared Folder**: NFS, SMB, or CIFS

- **Deduplication Storage Appliances**: ExaGrid, EMC DataDomain, HPe StoreOnce

The SOBR can expand with on-premises storage, such as block storage, or even a cloud-based object repository known as *Capacity Extent*. Veeam Backup & Replication combines the *Performance* and *Capacity* extents into one to summarize their capacities:

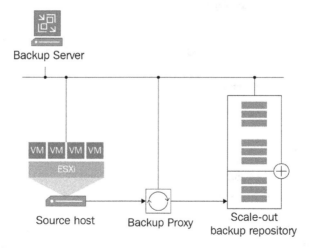

Figure 1.15 – Scale-out backup repository

The ability to use an SOBR is dependent on the license version that you are using with Veeam Backup & Replication:

- **Enterprise**: Allows for a total of two SOBRs with three extents
- **Enterprise Plus**: Provides for an unlimited number of SOBRs with as many performance extents as required, but only one capacity tier per SOBR

> **Tip**
> Should you happen to downgrade your licensing from Enterprise Plus or Enterprise to Standard, you will lose the ability to target your jobs to the SOBR. You can, however, restore data from the SOBR.

With the different license types, this limits you in terms of the number of both SOBRs and extents per SOBR you can configure. As noted above, there is a limit of two for *Enterprise*, and an unlimited number for *Enterprise Plus*.

> **Tip**
> For the best performance and manageability, it is best to keep your SOBR limited to 3-4 extents if possible. There is less movement of data and if you have a failed extent, replacing it is fairly simple. Also, when you have too many extents and need to move data around, this can become a bit of a challenge when extents start getting full. If you are using object storage, then it will be one of the components of the SOBR that will be the capacity tier.

The SOBR works with many types of jobs or tasks in Veeam Backup & Replication:

- Backup jobs
- Backup copy jobs
- VeeamZIP jobs
- Agent backups – Linux or Windows agent v2.0 or later
- NAS backup jobs
- Nutanix AHV backup jobs

The next thing to keep in mind is the limitations of using a SOBR as there are certain things you cannot do:

- Only Enterprise & Enterprise Plus License support using a SOBR.
- You cannot use it as a target for configuration backup jobs, replications jobs, VM copy jobs, Veeam Agents v1.5 or earlier for Windows, and v1.0 or earlier for Linux.
- Adding a repository as an extent to a SOBR will not be allowed if there is an unsupported job using the repository.
- Rotating drives are not supported – an example would be a drive attached by USB or serial cable.
- You are unable to use the same extent in two scale-out repositories.

Refer to the following page limitations on the Veeam Backup & Replication website: `https://helpcFixedenter.veeam.com/docs/backup/vsphere/limitations-for-sobr.html?ver=100`.

When it comes to the makeup of the scale-out, there are two tiers:

- **Performance Tier**: Fast storage and fastest access to data
- **Capacity Tier**: Typically object storage for long-term archival and offloading capabilities or a lower tier of storage

The performance tier you require will be the one that provides the quickest access to files and restores as and when necessary. When you create a standard repository before adding it to an SOBR, there are specific settings retained in the SOBR:

- The number of simultaneous tasks it can perform
- The storage read and write speeds
- Data decompression settings relating to storage
- The block alignment settings of the storage

What the SOBR will not inherit is a repository backed by rotating drives or if you selected to use the per-VM backup option. This option is on by default in a SOBR.

Another aspect to consider is the backup file placement policy that you will use. There are pros and cons to both, and specific operating systems such as ReFS and XFS require one over the other. The two types of placement policies are as follows:

- **Data locality**
- **Performance**

Refer to the following performance tier page on the Veeam Backup & Replication website: `https://helpcenter.veeam.com/docs/backup/vsphere/backup_repository_sobr_extents.html?ver=100`.

Data locality allows the scale-out to place all backup files in the chain to the same extent within the SOBR, thereby keeping files together. In contrast, the performance policy will enable you to choose which extents to use for both full backup files (VBK) and incremental files (VIB). The metadata file (VBM) is located on all extents in the SOBR for consistency and in the case of files needing to move extents:

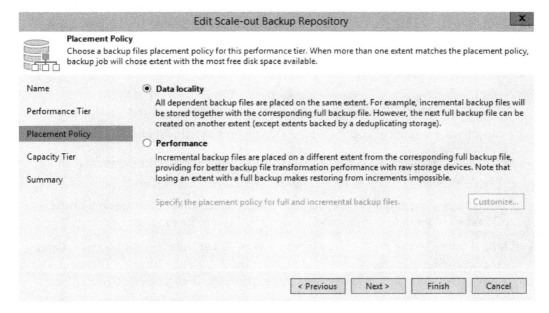

Figure 1.16 – Placement policy options for the scale-out repository

For further information on backup placement, refer to the following page on the Veeam Backup & Replication website: `https://helpcenter.veeam.com/docs/backup/vsphere/backup_repository_sobr_placement.html?ver=100`.

Now, when it comes to the capacity tier, there can only be one per scale-out, and it is required to be one of the following:

Object Storage
Select the type of object storage you want to use as a backup repository.

 S3 Compatible
Adds an on-premises object storage system or a cloud object storage provider.

 Amazon S3
Adds S3 object storage or an AWS Snowball Edge appliance.

 Microsoft Azure Blob Storage
Adds Microsoft Azure blob storage. Both Azure Blob Storage and Azure Data Box are supported.

 IBM Cloud Object Storage
Adds IBM Cloud object storage. S3 compatible versions of both on-premises and IBM Cloud storage offerings are supported.

Cancel

Figure 1.17 – Object Storage repository for the capacity extent

Using a capacity tier as part of your SOBR is suitable for the following:

- You can offload older data or when your SOBR reaches a specific percentage of capacity to allow you to free up storage space.

- Company policy stipulates that you have to keep a certain amount of data onsite, and that all older data is then tiered off after X days to the capacity tier.

- Using it falls into the 3-2-1 rule, where one copy of the information is offsite. Refer to the following blog post for more details on the 3-2-1 rule: https://www.veeam.com/blog/3-2-1-rule-for-ransomware-protection.html.

You specify the capacity tier after creating it as a standard repository and during the SOBR wizard at this stage:

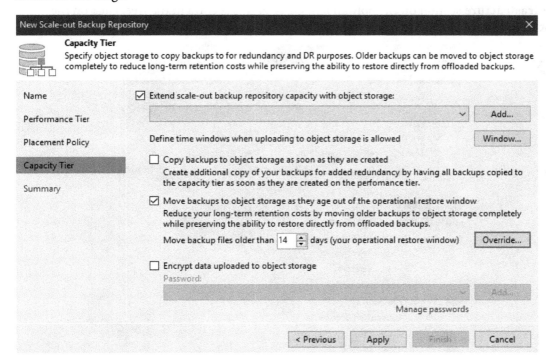

Figure 1.18 – Capacity Tier page of the scale-out wizard

Please visit the following capacity tier page on the Veeam Backup & Replication website: `https://helpcenter.veeam.com/docs/backup/vsphere/capacity_tier.html?ver=100`.

We are now going to tie all of the above information together and create an SOBR. First, you need to open the Veeam Backup & Replication console and select the **Backup Infrastructure** section on the bottom left. Once in this section in the tree, click on the **Scale-out Repositories** option on the left:

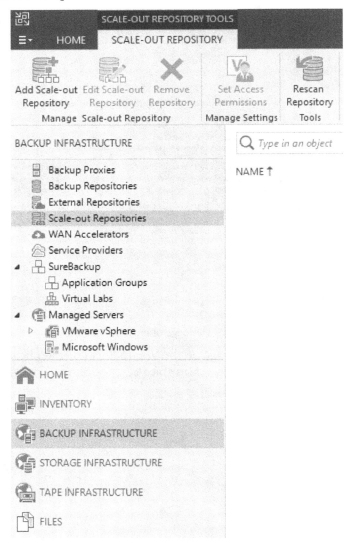

Figure 1.19 – The Scale-out Repositories section of the console

Once you are in this section, you can either click the **Add Scale-out Repository** button in the toolbar or, on the right-hand pane, you can right-click and select **Add Scale-out backup repository…**.

You will then name the scale-out and give it a thoughtful description. The default name is **Scale-out Backup Repository 1**. You then click **Next**, and you will come to the **Performance Tier** section of the wizard:

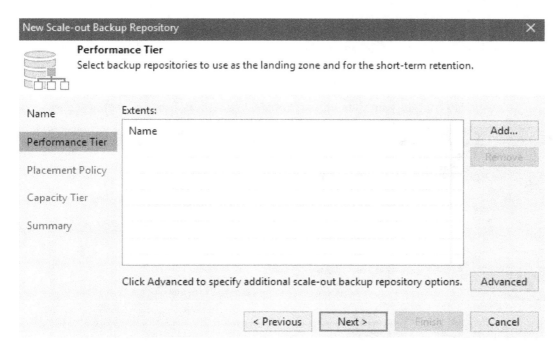

Figure 1.20 – Scale-out wizard – Performance Tier

In this section, you will click on the **Add...** button and choose the standard repositories that will be part of your scale-out. You can also click on the **Advanced** button to choose two options:

- **Use per-VM backup files** (recommended)
- **Perform full backup when the required extent is offline**

Click the **Next** button to proceed. You then will pick your placement policy, which is **Data Locality** or **Performance**. As noted, if using ReFS or XFS, you must select **Data Locality** to take advantage of the storage efficiency that each filesystem provides. Click **Next** after making your choice.

You can now choose to use the **Capacity Tier** option for your SOBR or just click the **Apply** button to finish. Note at this screen, when you select a capacity tier, that there are several options you can enable:

- Copy backups to object storage as soon as they get created in the performance tier.

- Move backups to object storage as they age out of the restore window – the default is 14 days, and you can also click on the **Override** button to specify offloading until available space is below a certain percentage.

- You can also encrypt your data upload to object storage as another level of security:

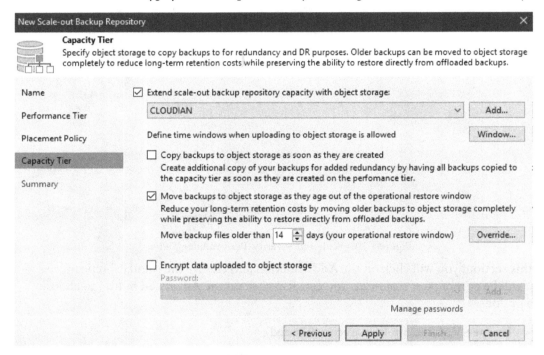

Figure 1.21 – Capacity tier selection for the scale-out

Note that some capacity tier targets support immutability. This feature is an essential attribute in the war on ransomware. In v10, capacity tier targets that support immutability include AWS S3 with object lock and some S3-compatible object storage systems. Immutability allows data to be protected and not deleted for the set period of days assigned.

Once complete, you will see your new SOBR, and when you select it, you will see the *performance tier extents* and *capacity tier* options if you chose it:

BACKUP INFRASTRUCTURE						
Backup Proxies	NAME ↑	TYPE	HOST	PATH	CAPACITY	FREE	USED SPACE
Backup Repositories	Backup Repository 1	Windows	home-vbk01.home.lab	F:\Backups	499.9 GB	496.8 GB	0 B
External Repositories	CLOUDIAN	S3-compatible		amazonS3://s3-...	N/A	N/A	0 B
▲ Scale-out Repositories						
Scale-out Backup Repository 1						

Figure 1.22 – SOBR created

If you want further information on the SOBR, visit this page on the Veeam Backup & Replication website: `https://helpcenter.veeam.com/docs/backup/vsphere/sobr_add.html?ver=100`.

Once set up within Veeam Backup & Replication, the SOBR is pretty self-sufficient. Still, there is maintenance that you need to perform in order to ensure optimal performance, and plenty of storage is available for backups.

The final thing to discuss is the management of the SOBR after creation. Once created, you may need to do any of the following items:

- Edit the settings to change the performance policy, for example.
- Rescan the repository to update the configuration in the database.
- Extend the performance tier by adding another extent to the SOBR.
- Put an extent in maintenance mode to either perform maintenance on the server that holds it or evacuate the backups to remove the extent.
- Switch an extent into what is called **Sealed Mode**, where you do not want any more writes to it, but you can still restore from it; this allows you to replace the extent with a new one.
- Run a report on the SOBR.
- Remove an extent from the SOBR that requires maintenance mode.
- And lastly, remove the SOBR altogether.

For more information on SOBR management, please visit the following Veeam Backup & Replication website: `https://helpcenter.veeam.com/docs/backup/vsphere/managing_sobr_data.html?ver=100`.

Summary

This chapter has provided you with the tools required to complete the installation of Veeam Backup & Replication v10 and outlined what components make up the installation. We discussed the pre-requisites, including versions of SQL Server that you can use – Express or SQL Standard/Enterprise. We next addressed in detail how to set up proxy servers, as well as configuration best practices and optimal settings. This part was then followed by a discussion on repositories and how to create them, and also included best practices and optimizations for best performance. Lastly, we looked at the scale-out repository and how to set it up, including the performance and capacity tiers, as well as how to manage them after setup.

This chapter will ensure that you have all the basics covered and will help when we move to the next chapter, which covers the 3-2-1 rule—*Chapter 2, The 3-2-1 Rule – Keeping Data Safe.*

2
The 3-2-1 Rule – Keeping Data Safe

When it comes to protecting your data, you must have more than one copy to ensure data safety. In this chapter, we will cover the **3-2-1 rule** and what it means. You will learn how to take this information and set up your backups based on this rule. You will also learn about the different types of media, such as **replication**, **snapshots**, and even **Capacity Tier Copy Mode (part of SOBR)**. If you adhere to the 3-2-1 rule, it ensures your data is safe, and at least one copy is offsite, which is where the 3-2-1 rule for backups comes into play:

- **3 copies**: Keep three copies of your data, one being the primary and two being the secondary copies.
- **2 media types**: Store your backups on at least two different types of media.
- **1 offsite**: Ensure that at least one copy of your data is kept offsite.

In this chapter, we're going to cover the following main topics:

- Introducing what the 3-2-1 rule is
- Understanding how you apply the 3-2-1 rule to backup jobs
- Exploring what media is best suited for use with the 3-2-1 rule

Technical requirements

One of the main requirements will be that you have **Veeam Backup & Replication** set up to go through some of the examples given in the following sections. If you have followed *Chapter 1*, *Installation – Best Practices and Optimizations*, then you have all the prerequisites required.

Guest perspective from Rick Vanover of Veeam

The 3-2-1 rule is an excellent mindset for *backup data management*. The beauty of this rule is that it doesn't require any specific type of hardware, yet it can address nearly any kind of failure scenario. Some scenarios can include *ransomware, hardware failure, loss of network/power/site*, and *accidental deletion*.

Keep the 3-2-1 rule in play as a minimum viable configuration for how many copies of data Veeam can manage. The recommendation is to have at least one copy of data that is either offline or air-gapped for the highest resiliency levels.

When implementing the 3-2-1 rule with Veeam, there is flexibility everywhere. It doesn't necessarily mean more backup jobs or data transfers are needed. Backup copy jobs, replica jobs, storage snapshots, SOBR Copy Mode, and more can all be ways to achieve the 3-2-1 rule.

Introducing what the 3-2-1 rule is

The 3-2-1 rule for backup is a reliable methodology to protect your data against a variety of things, including the following:

- **Ransomware**: Deleting or infecting backup files
- **Corrupted media**: Unable to read or restore from backups like tape media or disk
- **Data center loss**: When one copy is sent offsite elsewhere

The following diagram illustrates the 3-2-1 rule with backups to primary storage and secondary storage, and tape stored offsite, meeting the requirements:

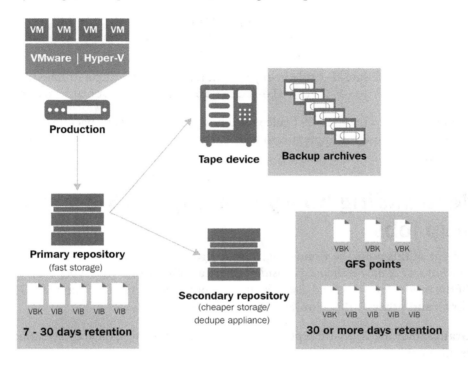

Figure 2.1 – Diagram of the 3-2-1 rule

As shown, following the 3-2-1 rule is easy when using different means of storage, which is further explained in the following section, covering the importance of the rule.

The importance of the 3-2-1 rule

The 3-2-1 backup strategy has been around for decades, but the principle remains the same. This backup strategy is recognized as best practice by Veeam and other information security professionals and government authorities. While this strategy doesn't mean that data is never compromised, it does eliminate many risks involved as it ensures there is no single point of failure for your data. It covers you in case one copy should become corrupted or if one of the many technologies you use fails. It can also cover you for theft if, say, ransomware wipes out your data.

One of the best things about setting up your backups for the 3-2-1 rule is the many resources online, one of them being on the Veeam blog:

- How to follow the 3-2-1 backup rule with Veeam Backup & Replication: `https://www.veeam.com/blog/how-to-follow-the-3-2-1-backup-rule-with-veeam-backup-replication.html`
- How Veeam and the 3-2-1 rule help you fight ransomware: `https://www.veeam.com/blog/3-2-1-rule-for-ransomware-protection.html`

We have now covered what the 3-2-1 rule is and how it plays into your backup strategy. We will now look at how you apply this to your backup jobs to ensure you are adhering to the 3-2-1 rule.

Understanding how you apply the 3-2-1 rule to backup jobs

So you might, at this point, be wondering, *how do I go about applying the 3-2-1 backup rule and what sort of media should I consider*? There are plenty of media types that you can use to fit the model for 3-2-1, and also take into account setting up retention with **GFS** (short for **Grandfather-Father-Son**), which breaks down as follows:

- **Grandfather**: Backups are kept for a year or more depending on your retention requirements.
- **Father**: Backups are kept for a month or more depending on your retention requirements.
- **Son**: Backups are kept for a week or more depending on your retention requirements.

You can set this up in the following manner:

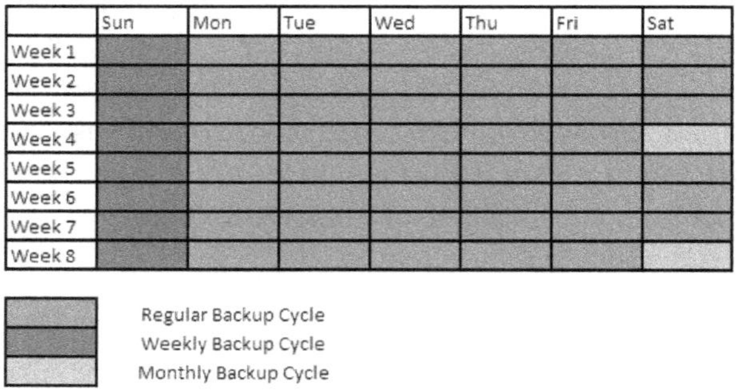

Figure 2.2 – Backup cycle example of GFS

When you are setting up your backup jobs within Veeam Backup & Replication, you will have the option to configure the number of days/restore points:

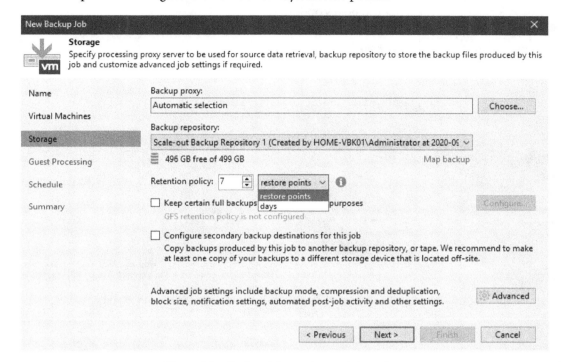

Figure 2.3 – Backup job – restore points or days

For the **Retention policy**, the following is an explanation of the drop-down options available:

- **restore points**: Veeam Backup & Replication keeps the last N restore points, where N equals the number of restore points.

- **days**: Veeam Backup & Replication will keep backups created for the previous N days, where N equals the number of days.

It is also within this configuration window that you set up your GFS retention policy, which works with the 3-2-1 rule when used in conjunction with the **Keep certain full backups longer for archiving purposes** option:

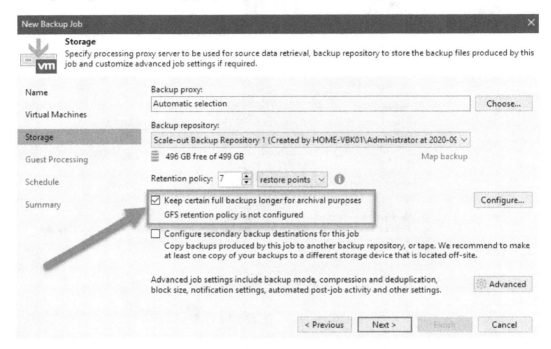

Figure 2.4 – GFS and secondary backup

When you select the GFS retention policy, you need to click the **Configure…** button to tell Veeam Backup & Replication how many backups of each type you want to keep and for how long:

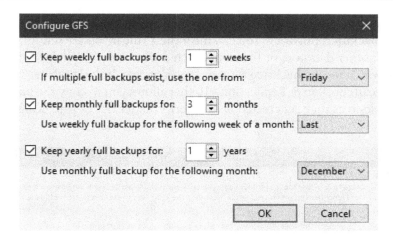

Figure 2.5 – The Configure GFS settings

Furthermore, before you can select the option to configure a secondary backup destination, you need to create either a backup copy job or a backup to tape job within Veeam Backup & Replication. This option fits the 3-2-1 rule, and when combined with another job type that sends data offsite to another form of block/object storage or to tape for taking offsite, you ensure that you have the *1* of the rule covered: *offsite*:

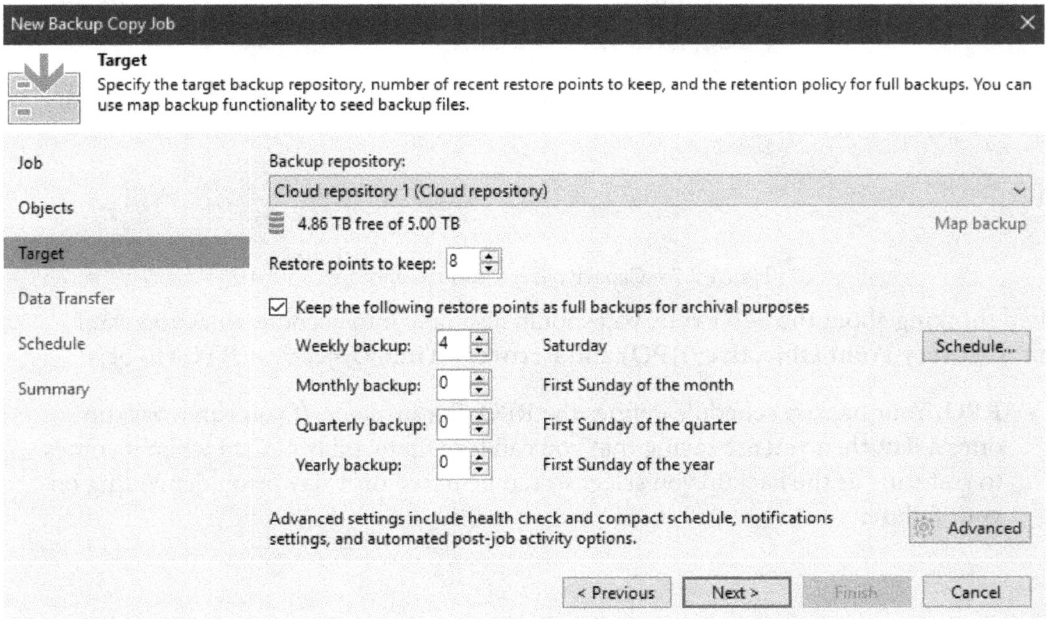

New Backup Copy Job

Target
Specify the target backup repository, number of recent restore points to keep, and the retention policy for full backups. You can use map backup functionality to seed backup files.

Job

Objects

Target

Data Transfer

Schedule

Summary

Backup repository:

Cloud repository 1 (Cloud repository)

4.86 TB free of 5.00 TB   Map backup

Restore points to keep: 8

☑ Keep the following restore points as full backups for archival purposes

Weekly backup: 4   Saturday   Schedule...

Monthly backup: 0   First Sunday of the month

Quarterly backup: 0   First Sunday of the quarter

Yearly backup: 0   First Sunday of the year

Advanced settings include health check and compact schedule, notifications settings, and automated post-job activity options.   Advanced

< Previous   Next >   Finish   Cancel

Figure 2.6 – Backup copy job for secondary copy

Another way to ensure you comply with the 3-2-1 rule would be by using the capacity tier of the **scale-out backup repository**. You can meet the *1* rule if you are sending it to an S3 target that is in another datastore or the cloud, such as Amazon S3. When sent to S3, that is on-premises, which meets the *2* rule for two copies of data, but not offsite. There is an option within Veeam Backup & Replication v10 that allows you to copy the backups to the Capacity Tier as soon as they get created on the performance tier of your scale-out backup repository:

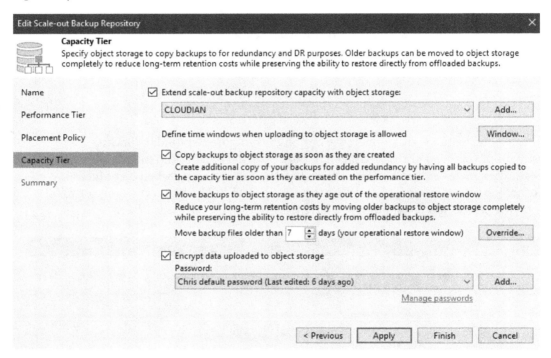

Figure 2.7 – Capacity Tier – immediate copy

When thinking about the 3-2-1 rule, you should also take into account what you want your **Recovery Point Objective (RPO)** and **Recovery Time Objective (RTO)** to be:

- **RPO**: Your backup schedule defines the RPO. For instance, if you run a backup once a day, then you are saying that you can lose up to 1 day of data when it comes to restoring as the backup you select would be based on 1 day prior, depending on your failure.

- **RTO:** The RTO is the amount of time it will take to restore your application and data to a running state. It is basically how long your business can be down without data. Is it 2 hours? 24 hours? 3 days? A week? Every business case will be different, so ensure you evaluate this with them before setting things up.

Regardless of the RTO and RPO settings, Veeam recommends having thorough conversations with stakeholders to ensure that the configuration of your backup implementation is in line with their expectations. Otherwise, there is potentially a gap between what gets implemented and what the expectations are. This discussion can also be an excellent catalyst for funding IT projects if investments in storage, networking, or software are needed to bridge the gap.

We have now learned how to apply the 3-2-1 rule when creating jobs by setting up specific jobs and options. We will look at the media that can be used with the 3-2-1 rule to meet all of the requirements.

Exploring what media is best suited for use with the 3-2-1 rule

When it comes to implementing the 3-2-1 rule for backups, you want to ensure that you follow the best practices, which means to ensure you use at least two different types of media. There are many media types out there that are suitable for use within Veeam Backup & Replication.

Disk media includes the following:

- SAN
- Storage snapshots (primary storage)
- NFS **Network-Attached Storage (NAS)** devices
- SMB NAS devices
- USB
- Rotating/removable media, such as RDX drives
- A general-purpose disk that is directly attached storage
- Replication jobs to primary storage for VMs (VMware vSphere and Microsoft Hyper-V)

Tape media includes the following:

- LTO
- WORM LTO
- **Virtual Tape Library (VTL)**

Deduplication appliances include the following:

- Dell Data Domain
- HPE StoreOnce
- ExaGrid
- Quantum DXi
- Other deduplication appliances presenting file shares
- Filesystems with deduplication technology (Windows' deduplication feature)

BaaS/DRaaS include the following:

- Backup as a Service
- Disaster Recovery as a Service

The cloud includes the following:

- Azure blob
- AWS S3
- S3-compatible
- IBM Cloud Object Storage
- Google Cloud Object Storage (coming in v11, announced at VeeamON 2020)

Depending on how you want to set up the 3-2-1 rule and your company requirements, you will choose media that best suits your scenario. We will look into the different permutations of backups with various media that meet the 3-2-1 rule as there are pros and cons to each, including some practical advice for each.

Scenario 1

1. Primary backup to disk – backup to a NAS or SAN with fast storage

2. Secondary copy to another form of a disk – backup copy to a deduplication appliance or lower-tier storage

3. Backup to tape – LTO tape backup

This permutation is one of the more typical scenarios that you will see in the backup industry, and it does meet the 3-2-1 rule and uses two forms of disk media, such as SAN and NAS:

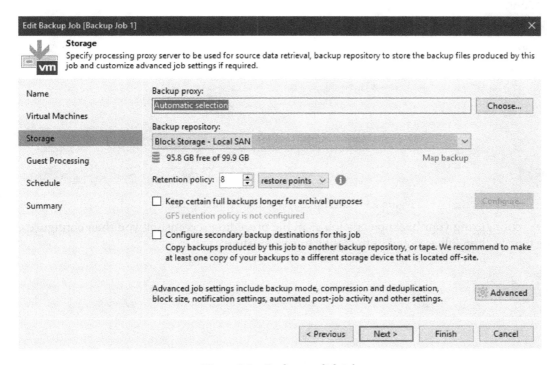

Figure 2.8 – Backup to disk job

The preceding screenshot shows setting your backups to go to your primary storage system, while the next one shows a backup copy job going to secondary disk storage:

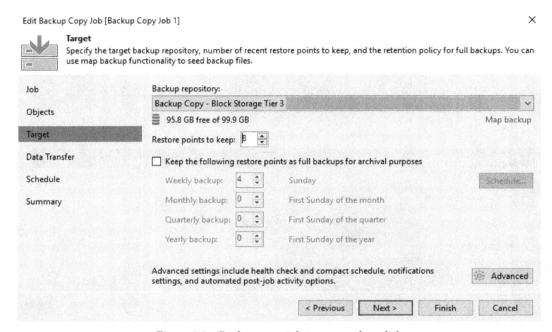

Figure 2.9 – Backup copy job to a secondary disk

After configuring your backup copy job as in the preceding screenshot, you then configure the tape backup job as noted in the following screenshot:

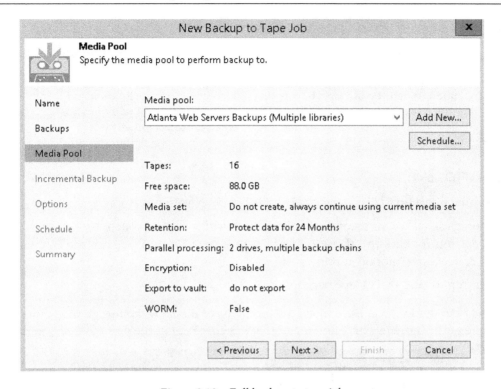

Figure 2.10 – Full backup to tape job

The pros are as follows:

- You meet the three-backups requirement using two forms of disk media as well as tape.
- Tape or the second disk could be the offsite copy.
- A tape backup is considered **air-gapped** media, which can prevent ransomware from infecting your data when sent offsite.

The cons are as follows:

- You are only using two forms of media, which could be susceptible to corruption.
- Tape media, if used for offsite storage, does take time to retrieve as you typically send it to a secure facility.
- When conducting a restore with tape media, it takes time to read the data from the tape.

> **Tip**
>
> When using this method for backup to meet the 3-2-1 rule, ensure that your first disk backup is to fast storage but only kept for a minimal amount of time. The backup copy to the secondary disk and tape media can hold longer retention. When you use tape, be sure that it is stored in a secure location and that you test restores frequently. Tape media can last for up to 30 years, but it is all dependent on where and how it is stored, which can affect longevity.

Scenario 2

1. Primary backup to disk – backup to a NAS or SAN with fast storage.

2. Storage snapshot for secondary – integration with a storage system for taking snapshots for backup either locally or, in the case of NetApp, you can use SnapVault to create a snapshot on another array.

3. Backup to tape – LTO tape backup.

With this permutation, you use a much different type of media with **storage snapshots**. This is one of the more recent technologies that has come out in the past 10 or so years:

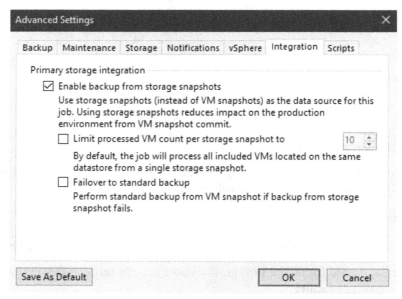

Figure 2.11 – Backup from storage snapshots integration

The pros are as follows:

- **Veeam Backup & Replication v10** has storage snapshot integration that can take backups directly from the snapshot, not affecting your production volumes or servers – an example of this would be with **NetApp storage**.

- **Veeam Explorer for Storage Snapshots** for recovery – this allows you to browse the storage snapshots to pick a point in time for the restore process.

- Both disk and storage snapshots are typically on much faster media.

- Restoring from both disk and storage snapshots is quick.

- Storage snapshots are easy to clone to a volume for testing a VM restore without affecting production.

- The tape would be the offsite copy.

- A tape backup is considered air-gapped media, which can prevent ransomware from infecting your data.

The cons are as follows:

- Not all storage supports storage snapshots, so you need to purchase more expensive storage arrays.

- Tape media, if used for offsite backups, does take time to retrieve as you typically send it to a secure facility.

- When conducting a restore from tape media, it takes time to read the data from the tape.

Tip

Using this method for backup to meet the 3-2-1 rule for both disk and storage snapshots will typically be high-speed media, so your *RTO* will be very low, as well as possibly your *RPO*. The storage snapshot integration is a great way to ensure that the impact on your environment will be minimal.

Please visit the following website for more information on the storage that integrates with Veeam Backup & Replication, as well as vendors:

`https://www.veeam.com/storage-integrations.html`

Scenario 3

1. Primary backup to disk – backup to a NAS or SAN with fast storage

2. **DRaaS** – VM replication to service provider – subscription to **Veeam Cloud Connect**

3. Backup to tape – LTO tape backup

While two of these backups are familiar, using a service provider for a DRaaS brings a whole new way to back up your data and servers using replication. Server or VM replication is a more recent technology that sends a "copy" of your server to the service provider, which sits dormant until you declare a disaster. At this point, it can be powered on and accessed:

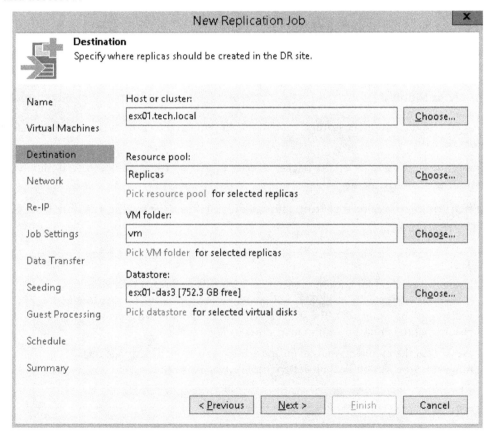

Figure 2.12 – Replication job to service provider

The pros are as follows:

- **VM replication** to a service provider ensures data resiliency as well as an offsite backup.

- Data center loss will not be an issue as you can power on your servers at the service provider.

- With VM replication, you can conduct test failovers to ensure your data is accessible and servers work from the service provider level.

- You can orchestrate failover testing using another product by Veeam called **Veeam Availability Orchestrator**.

  More information is available here: `https://www.veeam.com/availability-orchestrator.html`.

- Backup to disk is always fast and reliable if the **RAID** level is set up correctly and based on best practice, which is to use RAID 10 whenever possible (2x write penalty, but capacity suffers), and then RAID 5, with RAID 6 being the last choice.

  More information is available here: `https://www.veeam.com/blog/vmware-backup-repository-configuration-best-practices.html`.

- A tape backup is excellent for air-gap backup and offsite storage.

The cons are as follows:

- The service provider could have an outage that causes data loss and the need to reseed it again from your onsite backup.

- Connection to the service provider could go down due to various reasons, so your offsite replication would need to be paused.

- Disk failure depending on the setup could cause data corruption or loss.

- Tapes can wear out depending on how they are stored and other factors.

- Retrieval of tapes takes time, and so does data restoration.

> **Tip**
> If you are looking for a second offsite option other than disk or tape, using a **managed service provider** with DRaaS is a great alternative. Being able to replicate your entire servers and data over, then having the option to test failovers and use it live, allows you to ensure your data is safe. Also, service providers tend to have redundancy built into their solutions, which ensures that your information is secure.

Scenario 4

1. Primary backup to disk – **scale-out backup repository**
2. Capacity tier – **Copy Mode** with S3 object storage with object lock capability
3. Service provider – **Veeam Cloud Connect**

These three methods for the 3-2-1 rule use some of the more advanced features of Veeam Backup & Replication. The first backup option you have is the scale-out backup repository as your disk-based backup, which is made up of extents with the number dependant on your license. You then have the Capacity Tier within the scale-out backup repository with a copy mode that copies backups directly to object storage when completed on the performance tier. Veeam Cloud Connect allows you to create a backup job, a backup copy job, or a replication job depending on the services you get, thus providing another way to send your data offsite:

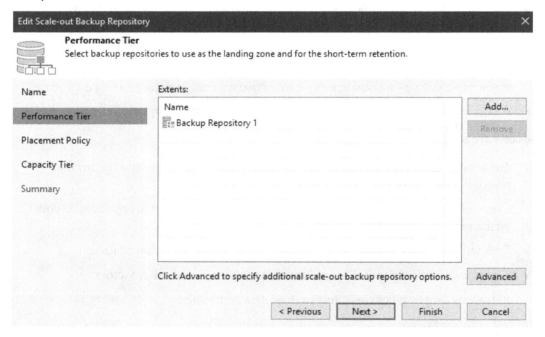

Figure 2.13 – Backup to disk – scale-out repository

The preceding screenshot shows creating a backup job using a scale-out backup repository, while the next screenshot shows the Capacity Tier of the scale-out using Copy Mode:

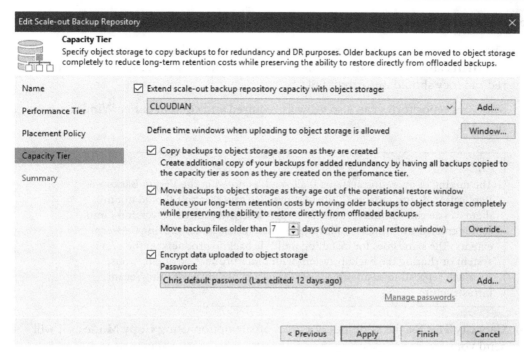

Figure 2.14 – Capacity Tier – Copy Mode turned on

Using the Capacity Tier allows you to use object storage such as **Cloudian,** as noted in the preceding screenshot, while configuring a service provider, as shown in the following screenshot, will enable you to use **Veeam Cloud Connect** for your backups:

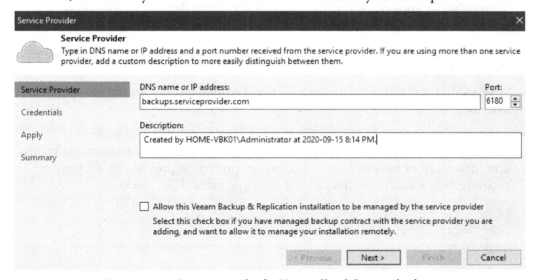

Figure 2.15 – Service provider for Veeam Cloud Connect backup

The pros are as follows:

- The scale-out repository offers resilience as you set up the performance tier or **extents** and can have more than one, allowing for disk space balancing and redundancy should one extent fail.

- Scale-out repositories can also grow as required and contain either Windows or Linux.

> **Tip**
>
> The mixing of operating systems is allowed, as noted in the Veeam Backup & Replication documentation. Still, best practice says that you should not mix them as you can run into problems. In this case, a Windows server and Linux server created with different filesystems could cause issues moving between extents. The same goes for installing multiple backup products on the same system or sharing the backup system with other production applications. As much separation as possible gives the best experience with the Veeam infrastructure.

- Capacity Tier for the scale-out is a great offsite option using **Copy Mode** as it will send your backups as soon as they complete on the performance tier when object storage is not on-premises.

- Depending on the vendor you use for object storage within the Capacity Tier, you can also have an **object lock** to prevent *ransomware* and make your data *immutable*.

- Immutable backups can also protect against accidental deletion or malicious administration in the Veeam Backup & Replication console.

- Using a service provider that offers **Veeam Cloud Connect** allows for offsite data storage using a variety of methods noted.

- Also, using a service provider, you tend to get redundant data protection with their systems.

- Many service providers also provide Veeam Cloud Connect with insider protection, which prevents deleting backups through accidental deletion, malicious administrators, or ransomware. While insider protection provides an excellent recoverability scenario, do not count it as a copy in the 3-2-1 model.

The cons are as follows:

- If not set up correctly when backing up to a scale-out repository, it will not provide the best performance.

- If any of your backup chains or data gets corrupted going to the disk-based backup in scale-out, then it will also move to the Capacity Tier in the same state, so in this case, turning on the **Health Check** option would overcome this.

- The service provider could have an outage or hardware failure that could cause data corruption and make it unusable.

- If you or the service provider have internet outages, then data backing up will need to be paused, and your RPO or Service-Level Agreement (SLA) window may lapse.

> **Important Note**
>
> All of the mentioned technologies use many of the more advanced features of Veeam Backup & Replication. Scale-out repositories offer scaling, resilience, and performance. Using these with your local backup is a great value proposition. On top of this, to combine scale-out with the Capacity Tier, you have a one-two punch for backing up your data both locally and offsite. When choosing a vendor, be sure to look into those that offer object lock and immutability. Service providers provide many reasonable plans for data backup using DRaaS, so using one allows two or more copies of your data to be offsite, which is always a good thing.

Scenario 5

- Primary backup to disk
- Backup to tape
- Service provider – Veeam Cloud Connect

This permutation is another one of the more typical scenarios you will see in the backup industry, and it does meet the 3-2-1 rule using a disk, tape, and service provider. Both the tape and service provider are considered offsite backups.

The following screenshot shows a backup job configured for a service provider repository as noted by **Backup repository: Cloud repository 1**:

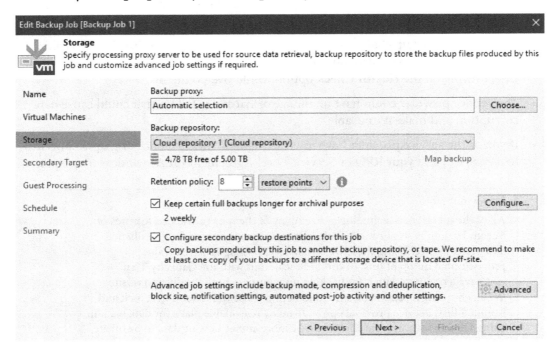

Figure 2.16 – Service provider backup job to cloud repository

The pros are as follows:

- You meet the three-backup requirement using three forms: disk, tape, and service provider.

- You will ensure data resiliency by having three forms of backup.

- Tape Backup and service provider Veeam Cloud Connect, both you can use for offsite copy.

- Tape backup is considered air-gapped media, which can prevent ransomware from infecting your data.

- Using a service provider service is one of the best ways to ensure your information is offsite and safe.

The cons are as follows:

- The service provider could have an outage that causes data loss and the need to reseed it again from your onsite backup.

- Connection to the service provider could go down due to various reasons, so your offsite replication would need to be paused.

- Disk failure depending on the setup could cause data corruption or loss.

- Tapes can wear out depending on how they are stored and other factors.

- Retrieval of tapes takes time, and so does data restoration.

> **Important Note**
>
> While using these three forms of media meets the 3-2-1 rule, there are other scenarios mentioned that are more resilient and offer even more data protection. When using disk-based backup, always ensure fast media for quick backups and restores and ensure you size according to restore points/days you want to retrieve with a shorter RTO. Always be sure you send tapes offsite, or if you keep them onsite, store them in a secure location. You will still want to test tapes to ensure you can restore them and meet your RPO. Also, when using a service provider for either a backup copy or replication, be sure to test restores often. Data rarely gets corrupted or deleted, but check just in case.

As you can see, there are multiple ways to set up your backup infrastructure to meet the 3-2-1 rule. There are also many types of media that you can use, and the different combination scenarios we covered demonstrated them.

Summary

This chapter reviewed the importance of keeping your data safe by following the 3-2-1 rule for backups. We took a look at what the 3-2-1 rule is and what it means when it comes to setting up your backups. Also covered was how this rule protects you from ransomware, corrupt data, and possible data center loss. We added in the GFS rule for backups to show how you can retain your data to meet your **SLA**, RPO, and RTO. Considerations for the different types of media you can use and various scenarios for backups were also given.

Both *Chapter 1, Installation – Best Practices and Optimizations*, and this chapter covered the basics with Veeam Backup & Replication for installation, including best practices and optimization and the 3-2-1 rule for data resiliency. The next chapter, on **NAS backups**, starts the next section of the book on the more advanced features.

Section 2: Storage – NAS Backup, Linux, SOBR, and OBS

The objective of this section is to show users different things regarding storage and NAS backup. The user will learn about the powerful new NAS backup feature of Veeam Backup & Replication v10 and the requirements for it. They will learn about using a Linux server as an XFS repository with Reflink and Fast Clone. We will even discuss object storage as part of a SOBR, along with the new Immutability features, including vendors that support Veeam. You will gain a wealth of knowledge about NAS backups and other options for scale-out repositories to use in your environments.

This section contains the following chapters:

- *Chapter 3, NAS Backup*
- *Chapter 4, Scale-Out Repositories and Object Storage – New Copy Policy*
- *Chapter 5, Windows and Linux – Proxies and Repositories*
- *Chapter 6, Object Storage – Immutability*

3
NAS Backup

Network-Attached Storage (**NAS**) is storage attached to your network in various forms, such as *Windows SMB*, *CIFS*, *NFS*, and *FreeNAS*, and provides a location for users to store data. A NAS can be set up in many different ways and allows faster data access, easier administration, and simple configuration over general-purpose file servers. This chapter discusses what NAS is and the most efficient way to back it up using Veeam Backup & Replication. We will discuss setting up NAS backup shares as well as setting up a NAS backup job. We will also discuss the components that Veeam Backup & Replication uses for the backup process. By the end of the chapter, you will have a better understanding of NAS backups.

In this chapter, we're going to cover the following main topics:

- Understanding NAS backup and what's new in version 10
- Learning how to configure NAS backup shares
- Discovering how to create NAS backup jobs
- Working with NAS backups – optimization and tuning
- Understanding the NAS restore options

Technical requirements

For this chapter, having Veeam Backup & Replication installed will be required, as well as having access to a NAS share for conducting backups. If you have followed along throughout the book, then *Chapter 1, Installation – Best Practices and Optimizations*, covered the installation and optimization of Veeam Backup & Replication, which you will leverage in this chapter. You can also reference the *NAS Backup* section of the Veeam website here:

```
https://helpcenter.veeam.com/docs/backup/vsphere/file_share_
support.html?ver=100
```

Understanding NAS backup and what's new in version 10

With the release of Veeam Backup & Replication version 10, the ability for users to protect a variety of NAS file shares was added. NAS can share data for many users on a network, from SMB to Enterprise. File shares come in a variety of setups:

- SMB v1, 2, and 3

- NFS v3, 4.1

- Windows File Server

- Linux File Server

- Enterprise NAS – Dell EMC Isilon, vSAN File Services

- NetApp – Storage Filer

Veeam Backup & Replication also uses changed file tracking, allowing speedy incremental backups. It is *snapshot-friendly*, allowing the orchestration of backups from snapshots created on an Enterprise-grade NAS device:

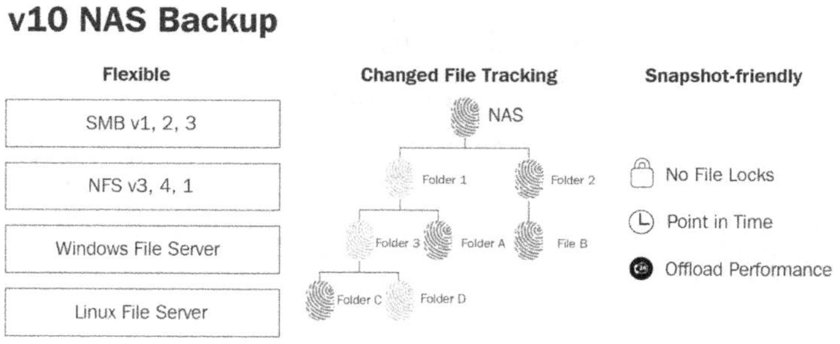

Figure 3.1 – NAS backup – file types, changed file tracking, and snapshots

Please note that we will look at the **Changed File Tracking** section of the preceding diagram in more detail in *Figure 3.3*.

The components that make up the backup for NAS are as follows:

- **The Veeam Backup & Replication server**: The primary server that delegates tasks to the file proxy servers
- **File proxies**: Similar to server backup proxies, which do the work
- **Cache repository**: Metadata repository and change file tracking location
- **Repository storage**: Where the backup data files are stored

The following diagram illustrates this:

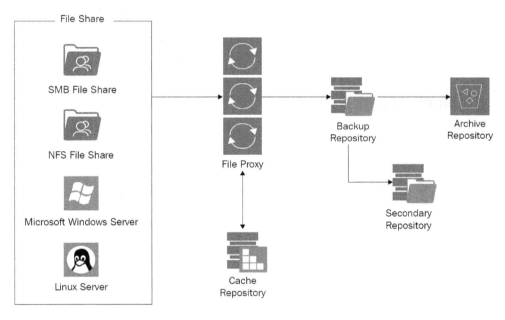

Figure 3.2 – Veeam Backup & Replication components for NAS backups

File proxy

The **file proxy** is the component that sits in between the file shares and the rest of the other elements of the backup infrastructure. File proxies are where job processing gets done, and the file proxy takes care of the backup and restore traffic on the network. Typically, this role is assigned to the backup server, but this is for small installations. A dedicated server should be set up for large Enterprise environments. For the optimal performance of many concurrent tasks, you will want to deploy several file proxies.

Cache repository

As you may have guessed, this is where metadata is stored and is used to reduce the load on the actual file shares themselves during backup. This location is where the **changed file tracking** takes place as Veeam Backup & Replication keeps track of all objects and the changes per backup session. It allows incremental backups to take place quickly as only file changes get backed up. The best performance for backup jobs is when the cache repository is close to the shares getting backed up:

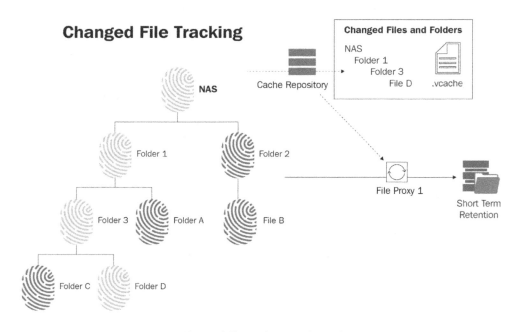

Figure 3.3 – Changed file tracking in the cache repository

Storage repositories

These are where the backup data is stored and retrieved from when a restore is required. As noted in *Chapter 2*, *The 3-2-1 Rule – Keeping Data Safe*, which covered the **3-2-1 rule**, there are many ways you can set up your repositories for backups. There could be the need for the archival of file backups depending on the type, so you need to consider this when configuring your repositories.

The required components can be referenced here, including a table of the different types of repositories:

```
https://helpcenter.veeam.com/docs/backup/vsphere/file_share_
support.html?ver=100
```

As you can see, **NAS backups** are flexible with the kind of shares they can back up, use change file tracking, and are snapshot-friendly. Next, we will cover setting up NAS shares in the Veeam console in preparation for backing them up.

Learning how to configure NAS backup shares

For you to back up your network shares, you need to add the network shares into the Veeam Backup & Replication server under the **INVENTORY** tab within the console:

Figure 3.4 – The INVENTORY tab where you set up file shares

When you click on the **ADD FILE SHARE** link shown in the preceding screenshot, it will open the **Add File Share** dialog shown in the following screenshot:

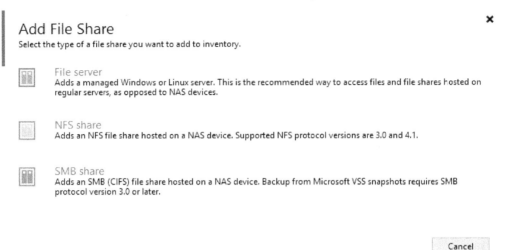

Figure 3.5 – The Add File Share dialog

This window is where you pick from the different types of file shares mentioned at the beginning of the chapter. There are the **File server** (Windows or Linux), **NFS share**, and **SMB share** options. Depending on your choice, you are presented with different wizards to go through:

- **File Server**: This wizard takes you through adding a managed server as the file server:

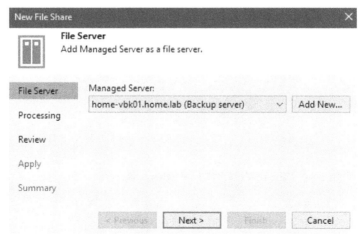

Figure 3.6 – File Server wizard

- **NFS File Share**: This wizard takes you through pointing to the NFS share located on your NAS device, including the path:

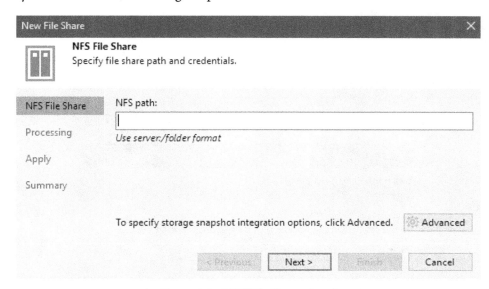

Figure 3.7 – NFS File Share wizard

- **SMB File Share**: This wizard takes you through pointing to the CIFS share located on your NAS device, including the path:

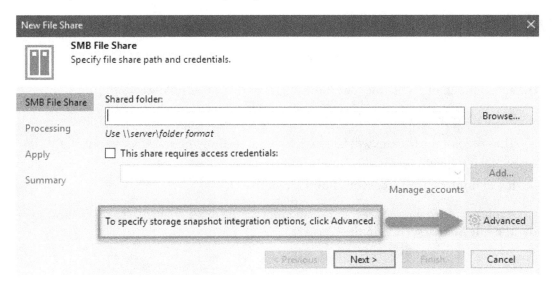

Figure 3.8 – SMB File Share wizard

The previous steps show you how each of the three different types of shares gets configured within the Veeam Backup & Replication console.

> **Important Note**
>
> Both the **NFS share** and **SMB share** options allow you to configure the **storage snapshot integration**, as discussed earlier in the chapter. Also, note that the SMB option can use access credentials for security to access the CIFS share.

The following diagram illustrates the storage snapshot option:

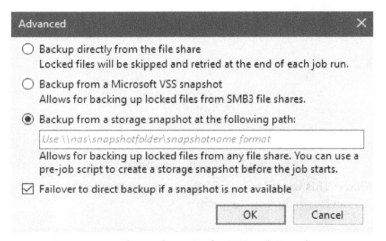

Figure 3.9 – Advanced options for NFS and SMB shares

Snapshot backup allows you to leverage the built-in snapshot functionality of the NAS device, where you schedule snapshots, and the backup job reads from the snapshot instead of the actual file share itself. The main advantages of this are as follows:

- The backup will not need to worry about locked or open files.
- The speed of the backup will be much faster at reading the snapshot versus the actual file share.
- It will improve your **Recovery Point Objective (RPO)**.

Once you have added your shares, and depending on the types, you will see them show up under the **File Shares** section in the console:

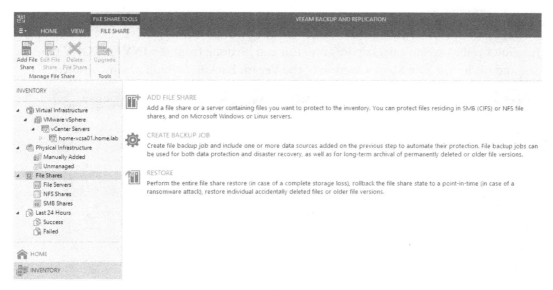

Figure 3.10 – File shares for file servers, NFS, and SMB

Now that we have covered how to add shares (*file server*, *NFS*, or *SMB – CIFS*) to the Veeam Backup & Replication server, in the next section, we will go through the scenarios to back up those shares.

Discovering how to create NAS backup jobs

Now that we have completed the configuration of the file shares, we will go through how to configure jobs for each of the different file share types:

- **File servers**: When you set up a file server for your organization and then add it to Veeam Backup & Replication, you can back up the entire server using **Microsoft VSS writer** snapshots.

- **NFS shares**: When you deploy a NAS device, you can set up NFS shares on it, which can then, in turn, be backed up.

- **SMB (CIFS)**: Also available on NAS devices when deployed, and you have the ability to back up.

To begin the backup process, take the following steps:

1. You need to create a backup job, and this can be done from the **INVENTORY** tab and under the **File Shares** section of the Veeam Backup & Replication console:

Figure 3.11 – The INVENTORY tab and the File Shares section of the console

2. From here, you select the **CREATE BACKUP JOB** option on the right side of the screen to launch the **New File Backup Job** wizard:

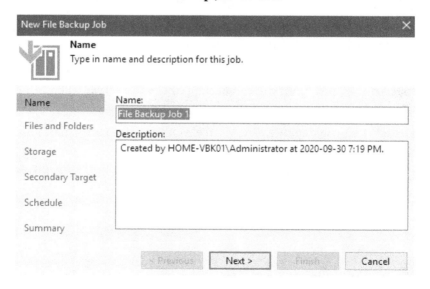

Figure 3.12 – New File Backup Job wizard

3. Name your job and enter a description, then click the **Next** button. On the next screen is where you will select from the different shares that you added to the **File Shares** section of the console by clicking the **Add…** button:

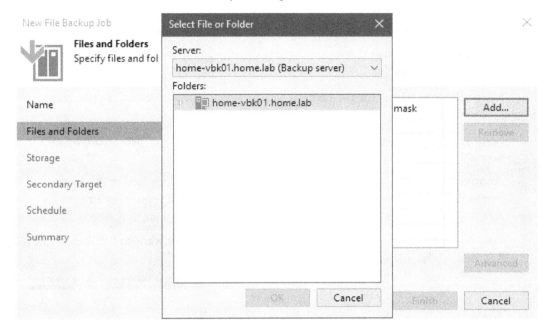

Figure 3.13 – Adding a file share to the backup job

4. From within the **Select File or Folder** popup, if you click on the dropdown under **Server**, you can see the various files shares listed:

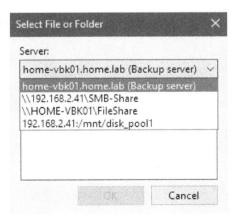

Figure 3.14 – File share selection for backup

5. From this drop-down list, you can pick the required file share for your backup job. If you have more than one, it is best to create separate backup jobs for each so that you can name each one appropriately. Once you have selected the share you want to back up and have clicked **OK**, you can then specify exclusions by clicking the **Advanced** button at the bottom of the screen. Click **Next** to continue.

6. The next part of the wizard is where you specify the repository you want to use, as well as other options:

    – **Keep all file versions for the last**: This is how many days or months to keep the files for backup.

    – **Keep previous file versions for**: This will keep active and permanently deleted files after they fall outside the retention policy specified.

    – **Archive repository**: This is where you will keep the files specified in the **Keep previous file versions for** option and this can be any local repository or also an archive, such as object storage or another S3 service (Cloudian).

    – **Files to archive**: This allows you to specify which files you want to archive, out of **All** files, **Active file versions**, and **Deleted file versions**:

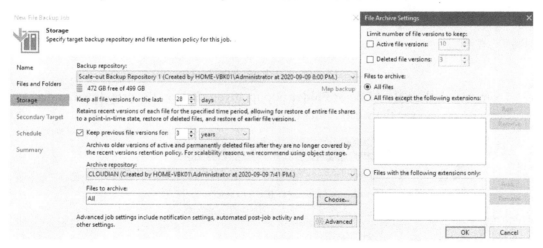

Figure 3.15 – Repository selection and file versions

7. If you click on the **Advanced** button, you can set options such as the following:

   – **ACL Handling**: How you want to back up permissions and attributes.

   – **Storage**: Compression level and encryption.

   – **Maintenance**: Set a file health check every so often – the latest backup file in the chain is checked to ensure consistency and the ability to restore from it by doing a **Cyclic Redundancy Check** (**CRC**) check on the metadata and a hash check on the data blocks to verify integrity. If that fails, a new full backup will be taken and a new chain started.

   – **Notifications**: Set up SNMP and email notifications for your job.

   – **Scripts**: Pre- and post-job scripts if required – where you can use a script to create a VSS snapshot before running the job and then another script to delete the VSS snapshot after the job completes:

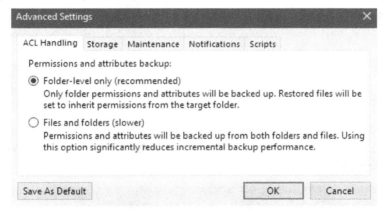

Figure 3.16 – Advanced features of the backup job

8. After you have clicked **Next**, the following screen allows you to specify a secondary target for your backup, a short-term file store for redundancy. Click **Next** to continue.

   When it comes to storage for file backup jobs, there are three types:

   a) **Backup repository**: This is the primary storage location for the backups and keeps backup files for the specified retention period.

   b) **Archive repository**: This is where object storage can keep backup files for a much longer retention cycle. Once files are copied to the archive, they are deleted from the primary backup repository.

c) **Secondary repository**: This will store a second copy of the backups from the primary backup repository by copying all backup files. The secondary can have its own retention and encryption settings. This setting is not configured by default and requires the user to enable it:

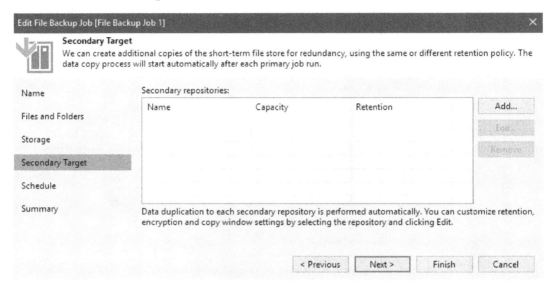

Figure 3.17 – Secondary target for short-term redundancy

9. After you add a secondary target, you can click **Edit...** to change the retention policy to something short term:

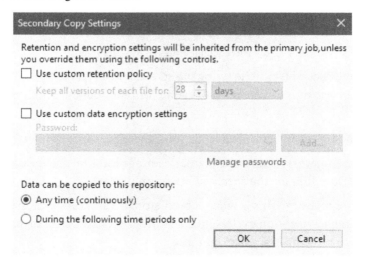

Figure 3.18 – Secondary target retention settings

10. The next screen allows you to set a schedule for your job, including **Automatic retry** attempts and the ability to terminate if it exceeds a set backup window time. It also allows you to override the retention policy set by the primary job. Click **Apply** to complete the job setup.

We have now completed adding file shares and created a backup job to back up our file shares. We will now look at working with the backup jobs created to optimize and fine-tune them.

Working with NAS backups – optimization and tuning

We have gone through adding **file shares** to Veeam Backup & Replication and also the process for setting up backup jobs. Now, we will look at the backup jobs created to work with them, including optimization and tuning.

When working with NAS backups, it is good to understand how the file share backup process works:

- When a backup session starts, Veeam Backup & Replication assigns a file proxy to process the file share information.
- The file proxy enumerates file and folder data, and a CRC tree is created.
- The file proxy transfers the CRC tree over to the cache repository.
- The cache repository saves the CRC tree.
- The file proxy now reads new data from the file share.
- The file proxy creates the backup data packages and transfers them to the target backup repository selected in the job.

- Veeam Backup & Replication checks files and if there was an archive repository configured in the backup job and retention was met, data is moved off to the archive repository:

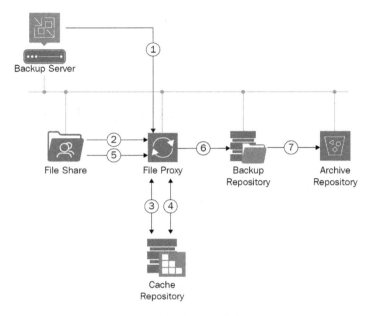

Figure 3.19 – File share backup process

For a great example of **retention scenarios**, please refer to the following Veeam Backup & Replication website and examples, as follows:

```
https://helpcenter.veeam.com/docs/backup/vsphere/how_file_
share_backup_works.html?ver=100
```

Example 1:

Only one file version is created, and the file does not change. File version 1 remains in the backup repository.

Example 2:

Retention of the backups is set to 5 days, the file changes daily, and one backup is performed each day. On day 6, file version 6 is added to the backup repository, and file version 1 is deleted by retention.

Whether you are adding a file server, NFS share, or SMB (CIFS) share, you have the option to use a cache repository as well as adjust the **Backup I/O control** setting to optimize the file share for backups:

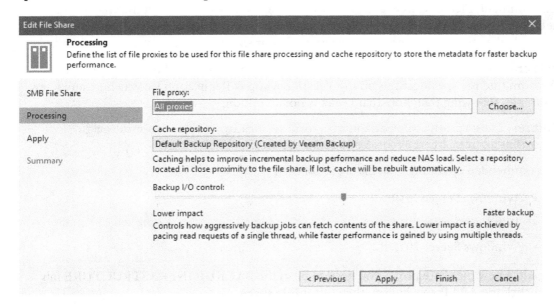

Figure 3.20 – File share optimizations for the cache repository and Backup I/O control

Making sure to make the necessary adjustments for this part of your file share will ensure that you achieve optimal performance. Selecting which cache repository to use helps the performance of the backup and reduce NAS load, as indicated. Also, moving the slider for **Backup I/O control** will help you achieve faster backups the further you move it to the right, including multithreading, whereas moving to the left sets slower reads with a single thread.

Tip

As noted in the screenshot, selecting a cache repository helps improve incremental backup performance and reduces the load placed on your NAS device or file server – this requires you to choose which cache repository you want to use. Also, adjusting **Backup I/O control** allows you to select how fast the file shares will get backed up; for example, if you schedule the backups during the evening, you may want it to be faster.

Depending on how many file shares and the amount of data you will be backing up, another optimization to look at is the file proxies and the number of them you have deployed. The more file shares and data, the more file proxies you will need in your environment. Please reference the *Further reading* section at the end of the chapter for a link to a great NAS backup calculator that allows you to enter variables. It will help you calculate the number of proxy servers and your sizing requirements.

File proxy servers also have specific requirements for hardware and operating systems. You can find the specifications on the Veeam Backup & Replication website here and a few examples following that; note that Linux is not an option:

```
https://helpcenter.veeam.com/docs/backup/vsphere/system_
requirements.html?ver=100#file_proxy
```

Recommended operating systems are as follows:

- Windows Server 2019

- Windows Server 2016

- Windows Server 2012 R2

To add a new file share proxy, you need to be in the **BACKUP INFRASTRUCTURE** tab and click the **Add Proxy** button, then select **File share...**:

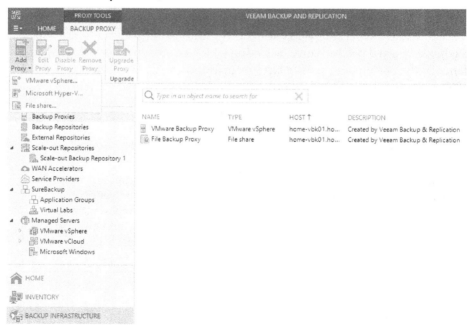

Figure 3.21 – Adding a file share proxy

In the **New File Proxy** wizard, you will select a Windows server from the drop-down list, or click the **Add New...** button to add a new server. You then choose the maximum concurrent tasks based on the number of CPUs that the server has. While there is no set rule on the number of CPUs that a file proxy should have, you should adhere to the same as a backup proxy server with a *maximum* of 8 and add more proxies as needed:

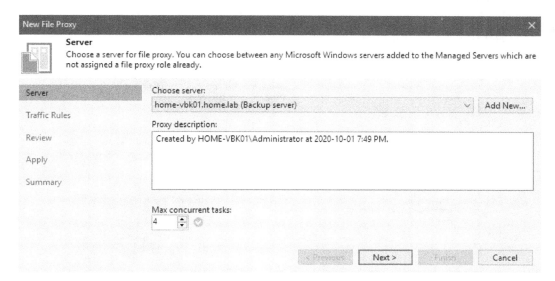

Figure 3.22 – New File Proxy wizard

In this section, we have learned the best ways to optimize and tune both our file shares and backup jobs. In the last section, we will cover the all-important NAS restore options.

Understanding the NAS restore options

After you have completed the setup of the NAS shares and the backup jobs, you will need to know how to *restore* from the backups. This section will discuss three options:

- **Restore entire share**: This will fix the whole share in the case of complete loss.

- **Rollback to a point in time**: This will allow you to select a snapshot from the backup of a date and time to revert to.

- **Restore individual files and folders**: You can restore an entire file or folder using this method.

If you were to lose your data center or servers, then you would need a way to restore your file shares. Then, you would use the option to restore the entire share either to the same location or an alternate location, including security and permissions.

To do this, you need to run the **Restore from File Backup** wizard and select the **Restore entire share** option:

Restore from File Backup

✖

Select the type of restore you want to perform.

 Restore entire share
Restores the latest version of all files to the selected location. Use this option in case of a complete loss of storage service, or major storage-level corruption impacting unknown number of files.

 Rollback to a point in time
Reverts all files modified since the specific date and time to the previous version, and restores all files that were deleted. Use this option to recover from ransomware, virus or insider attack.

 Restore individual files and folders
Restores the required file version, or point-in-time state of a folder to the specified location. Use this option to find and restore missing files or folders, or fetch previous file versions.

Cancel

Figure 3.23 – Restore from File Backup wizard – Restore entire share

Now, to use the **Rollback to a point in time** feature, you would select the second option, as noted in *Figure 3.23*, and then select the required date and time of the restore:

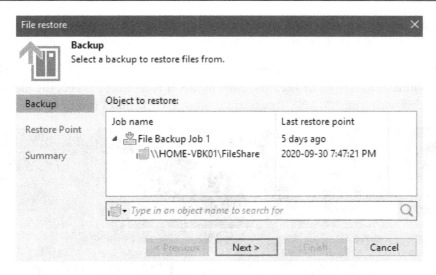

Figure 3.24 – Rollback to point in time selection

Lastly, to use the **Restore individual files and folders** option, you would select the third option, as noted in *Figure 3.23*. You then select the restore point required, as seen in *Figure 3.24* for the point-in-time restore. Once selected, click **Next** and then **Finish** to have the **BACKUP BROWSER** window open:

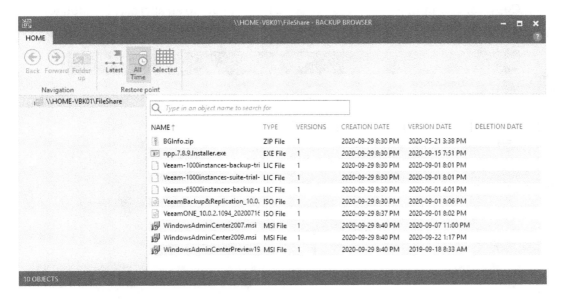

Figure 3.25 – The BACKUP BROWSER window for file or folder selection

With the **All Time** option selected, this will show you all the restore points in a single view. You can now right-click on any file or folder and select from two options:

- **Restore**: This allows you to restore the original location of the file or folder with the options to overwrite or keep the current file. If you select to keep, the file gets appended with _RESTORED_OriginalDate_OriginalTime:

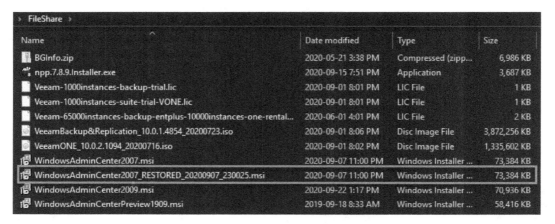

Figure 3.26 – Restored file when keep is selected

- **Copy To**: This option allows you to copy a file or folder to an alternate location for restore by first selecting the restore point and then the location:

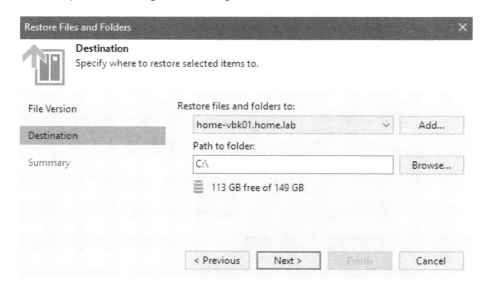

Figure 3.27 – Copy to location selection

You will now have a better understanding of the NAS Restore options available and how to use them within Veeam Backup & Replication.

Summary

This chapter reviewed NAS backups and what it takes to back them up. We took a look at what NAS stands for – **Network Attached Storage** – and how users connect to file shares, those being NFS, SMB (CIFS), or the use of file servers. We covered how to add a file share to the Veeam Backup & Replication console based on the type. Also covered was how to create backup jobs to then back up your file shares. We also took a look into optimizations as well as how you go about restoring your NAS backups.

Hopefully, you will have a better idea of what NAS backups are all about as the next chapter, *Chapter 4, Scale-Out Repositories and Object Storage – New Copy Policy*, will take a deep dive into the more advanced repository features.

Further reading

- NAS backup calculator: `https://cloudoasis.com.au/nas-calculator/`

4

Scale-Out Repository and Object Storage – New Copy Policy

Veeam Backup & Replication has a backup storage technology called **Scale-Out Backup Repository** (**SOBR**), used as a backup target for jobs. You can also attach *object storage* to a scale-out repository to further extend your backups to achieve longer-term retention. In this chapter, we will walk through how the scale-out repository gets created and how to use it best. We will also look at adding object storage repositories for the *Capacity Tier* of a scale-out repository. You will learn how to set up a scale-out repository, configuring it for performance and attaching object storage for the capacity tier. We will dive into the new *copy policy* for the capacity tier and how this plays a role in your backups. By the end of this chapter, you will know how to configure a scale-out repository with a capacity tier and set up the new copy policy to move backups out of the performance extents to object storage.

In this chapter, we're going to cover the following main topics:

- Understanding scale-out repositories and what is new in version 10
- Configuring standard repositories for use in a scale-out repository

- Setting up a scale-out repository
- Setting up an S3 object storage repository for use with a scale-out repository
- Configuring Capacity Tier for a scale-out repository, including the new copy policy

Technical requirements

For this chapter, installing Veeam Backup & Replication will be required, along with storage for creating repositories. If you have been following along through the book, then *Chapter 1, Installation – Best Practices and Optimizations*, covered the installation and optimization of Veeam Backup & Replication, which you will leverage in this chapter. You can also reference the *Scale-Out Backup Repository* section of the Veeam website here:

```
https://helpcenter.veeam.com/docs/backup/vsphere/backup_
repository_sobr.html?ver=100
```

Understanding scale-out repositories and what is new in version 10

Within the Veeam Backup & Replication software, there are two types of repositories where you store your backups:

- **Standard repository**: This can be **direct-attached storage (DAS)**, **Network File System (NFS)**, **Server Messaging Block (SMB)**, **Common Internet File System (CIFS)**, and Windows or Linux file servers, as well as deduplicating storage appliances, such as ExaGrid, Data Domain, and HPE StoreOnce:

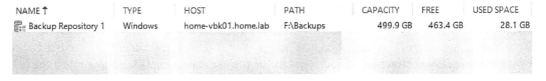

NAME ↑	TYPE	HOST	PATH	CAPACITY	FREE	USED SPACE
Backup Repository 1	Windows	home-vbk01.home.lab	F:\Backups	499.9 GB	463.4 GB	28.1 GB

Figure 4.1 – Standard performance repository in Veeam Backup & Replication

Also, adding a capacity tier is regarded as a standard repository as well, which you can use in the scale-out repository:

Figure 4.2 – S3-compatible standard repository

- **Scale-out repository**: This consists of standard repositories called *extents* and may also contain a capacity tier:

Figure 4.3 – Scale-out repository with performance and capacity extents

The scale-out repository gives you the ability to scale support for different tiers of data horizontally. It consists of one or more backup repositories called *extents* and can expand using on-premises or cloud-based object storage called *capacity extents*. Everything in a scale-out repository is joined together into a system, the capacity is gets summarized for the entire scale-out repository.

Some of the main benefits of the scale-out repository are as follows:

- Easy management of backup storage.

- The ability to scale as required by adding more capacity with new extents.

- It supports many backup targets, such as Windows and Linux servers with local or **Direct Attached Storage (DAS)** network shares and *deduplicating appliances*.

- It allows granular performance policy setup – **Data locality** or **Performance** modes.

- It provides unlimited cloud-based storage using the capacity extent for offloading data for long-term retention.

With Veeam Backup & Replication version 10, there are some much-needed enhancements to the scale-out repository, including the following:

- **Seal Mode**: You can now place extents in **Seal Mode**, which stops backups writing to them, but still allows restores and retention to take place. This option is also useful when you need to decommission an old server from the scale-out as it lets the data age out and then you can remove the extent.

- **Backup placement enhancements**: Both SQL Server and Oracle transaction log backups are now treated as incremental backups and placed on extents designated for incremental backups when using Performance mode.

- **Improved disk space reservation logic**: Extent disk space reservations get calculated using the full and incremental backup sizes for the job versus the static percentage of the machine's size. This will help with extent scheduling and when nearing capacity.

- **Tiering job status**: Now, when object storage moves and copy jobs occur, they have a **Waiting** status to help with troubleshooting. This status also means that there are tasks to process, but no backup repository slots are available.

- **Copy Mode for Capacity Tier**: When you add a capacity tier to your scale-out repository, there is an option to copy all backups to the object storage as soon as they get created.

You can reference these enhancements and many others on the Veeam Backup & Replication website:

```
https://www.veeam.com/whats-new-availability-suite.html
```

We have now covered what SOBR is and what makes up the components that you use to configure an SOBR. We will now take a look at the first part of a SOBR, that being the extent or standard repository configuration. This is the primary and only requisite component in order to be able to perform backups.

Configuring standard repositories for use in a scale-out repository

The fundamental piece of the scale-out repository is the *Performance* extent, which is built by adding *standard* repositories. When creating a standard repository, you need to decide on the type you will use, whether it be *Windows (ReFS)* or *Linux (XFS)*. You should use one or the other but not both together. While the documentation indicates that you can mix the repository types, most will agree that you use only one kind per scale-out for ease of management as well as having the same filesystem and performance. These filesystems use a newer technology called **Fast Clone**, which increases the speed of synthetic backup creation and transformation, reduces disk space, and decreases the load on storage. Please refer to the Veeam Backup & Replication website for more information:

```
https://helpcenter.veeam.com/docs/backup/vsphere/backup_
repository_block_cloning.html?ver=100
```

To create a standard repository, you need to set up your Windows or Linux server first using *ReFS* or *XFS* for the drive that will contain the backup data. When setting up your repository servers, there are considerations for each operating system:

1. **Windows (ReFS)**: Consider putting the data drives on their own VMware Paravirtual SCSI controller so as not to have the same traffic for the operating system on the same controller, plus paravirtual controllers have much better performance. Also, when formatting, use *ReFS* with 64K block sizing.

2. **Linux (XFS)**: Consider using *Ubuntu 20.04* as this has the latest kernel version, which Veeam recommends for use with repositories. Use this command, replacing /dev/sdb with the drive in your system to format the drive for data – mkfs.xfs -b size=4096 -m reflink=1,crc=1 /dev/sdb. Here, -b size=4096 is the system block size, reflink=1 enables reflinking for the XFS instance, and crc=1 enables checksums, required for reflink=1.

3. After setting up your repository server, you then add it to Veeam Backup & Replication by starting the **Add Backup Repository** wizard:

Add Backup Repository   ×

Select the type of backup repository you want to add.

 Direct attached storage
Microsoft Windows or Linux server with internal or direct attached storage. This configuration enables data movers to run directly on the server, allowing for fastest performance.

 Network attached storage
Network share on a file server or a NAS system. When backing up to a remote share, we recommend that you select a gateway server located in the same site with the share.

 Deduplicating storage appliance
Dell EMC Data Domain, ExaGrid, HPE StoreOnce or Quantum DXi. If you are unable to meet the requirements of advanced integration via native appliance API, use the network attached storage option instead.

 Object storage
On-prem object storage system or a cloud object storage provider. Object storage based repositories can only be used for Capacity Tier of scale-out backup repositories, backing up directly to object storage is not currently supported.

Cancel

Figure 4.4 – The Add Backup Repository wizard

4. Once the wizard launches, you then select the **Direct attached storage** option, where you choose either your Windows or Linux repository server:

Figure 4.5 – Direct Attached Storage selection for Windows or Linux

5. You make your choice, and then you are presented with a **New Backup Repository** window:

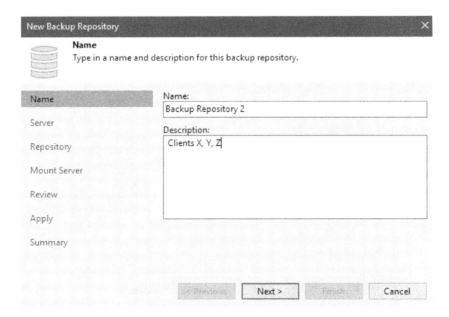

Figure 4.6 – The New Backup Repository wizard

Whether you selected *Windows* or *Linux*, this dialog is the same for both, but when you get to the next step for the server, that is where there are slight differences after clicking the **Add New** button:

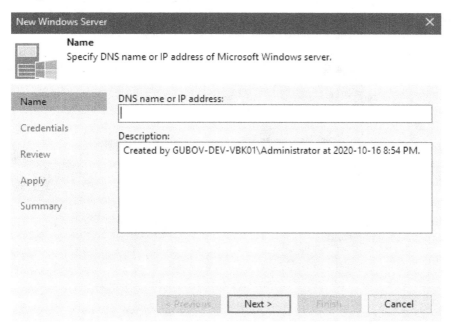

Figure 4.7 – Adding a new Windows server for a repository

The following dialog is for a Linux repository server:

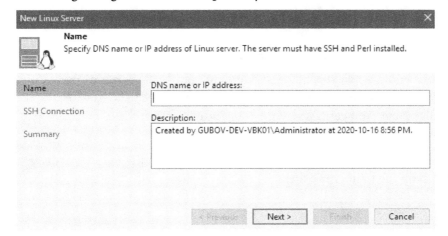

Figure 4.8 – Adding a new Linux repository server

As you can see, the title of each dialog relates to the server type you chose to add – **New Windows Server** or **New Linux Server**.

6. Once you have gone through the **Add New Server** wizard, you are then brought back to the **New Backup Repository** wizard, where you click the **Populate** button to show the drives on the server to select as the repository:

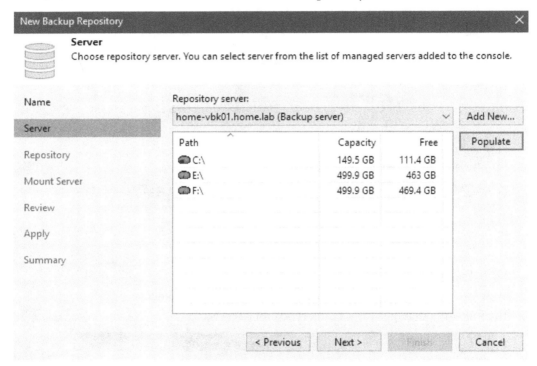

Figure 4.9 – Repository drive selection

7. As you can see, when you click the **Populate** button, the drives of the Windows or Linux server will appear for selection. Choose the drive to be used for the repository and click **Next**. You get taken to the **Repository** section of the wizard, where you set the various options:

   a) **Path to folder**: This is where you create a folder on the repository drive to store the backup job folders and files.

b) **Load control**: This has some settings that control the number of concurrent tasks to the repository and the read and write data rate. Both of these influence the load based on the type of repository you select and the backend storage system:

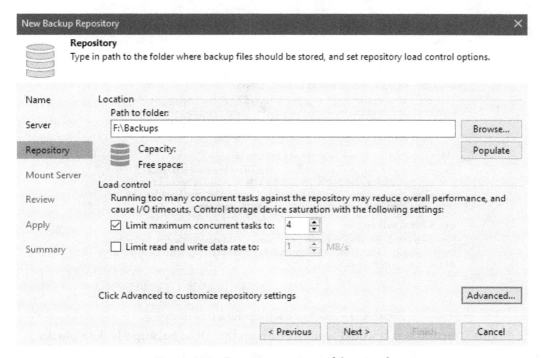

Figure 4.10 – Repository settings of the wizard

8. When you click the **Advanced...** button, you get the following dialog with more settings:

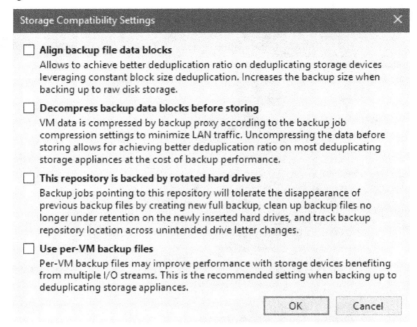

Figure 4.11 – Advanced settings for a repository

In this screenshot, you would typically turn on the **Align backup file data blocks** and **Decompress backup data blocks before storing** options when using a deduplication appliance such as the EMC Data Domain. **Use per-VM backup files** is also another good option for deduplication appliances, but is also recommended when using *ReFS* and *XFS* filesystems due to their performance.

9. After clicking **Next**, you are at the **Mount Server** section of the wizard where you select the following options:

    a) **Mount server**: The server used during the restore process that mounts the backup files.

    b) **Instant recovery write cache folder**: When you select to make an instant VM recovery, this folder mounts the files and typically gets placed on *SSD* or other high-performance storage.

c) **vPower NFS**: This is typically installed on the *mount server* and is for the instant recovery of any backup (physical, virtual, or cloud) to VMware vSphere. This service is not used for the *Microsoft Hyper-V* instant recovery options:

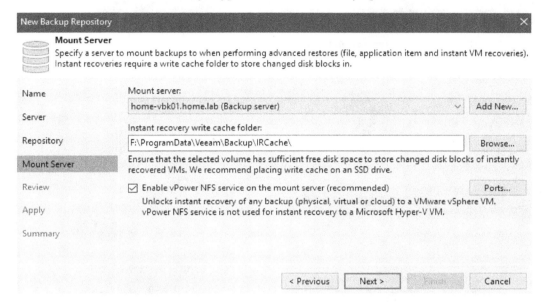

Figure 4.12 – Mount server settings

10. Once you select all the required options and click **Next**, you get prompted with the **Review** screen. Here, there are a couple of options to note:

   a) **Search the repository for existing backups and import them automatically**: When you choose a drive that may have been used previously as a repository, this option will search the drive and import the backups into the database.

b) **Import guest file system index data to the catalog**: This option will import any guest filesystem indexes that may have existed on the repository drive previously:

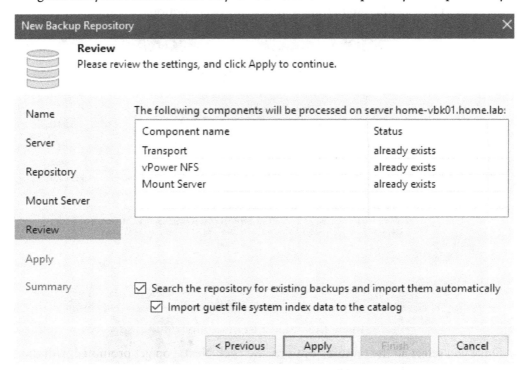

Figure 4.13 – Additional options on the review screen

11. Once you have reviewed everything, you click the **Apply** button, **Next**, and then **Finish** to complete the wizard.

This section completes the process to add a standard repository to the Veeam Backup & Replication console, which we will now use in the next section when configuring a scale-out repository.

Setting up a scale-out repository

In the previous section, you walked through setting up a standard repository, and we will now look at how you use this when creating a scale-out repository:

1. To walk through the scale-out repository, you select the **BACKUP INFRASTRUCTURE** tab and then click on **Scale-out Repositories**:

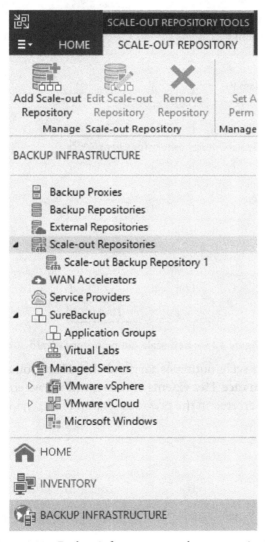

Figure 4.14 – Backup infrastructure scale-out repositories

2. To begin adding the scale-out repository, you can either click the **Add Scale-out Repository** button in the toolbar or right-click on the right-hand pane and select the same-named option. The **New Scale-out Backup Repository** wizard will launch, and you begin by giving it a name and a description, and then click **Next**:

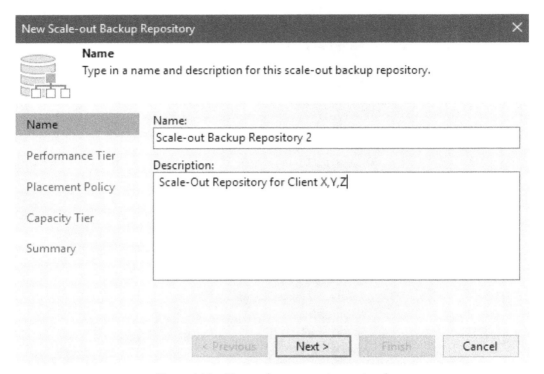

Figure 4.15 – New scale-out repository wizard

3. After you name your scale-out repository and click **Next**, you are then prompted to select the **Performance Tier** extents of the repository, where you will use the standard repository created in the previous section of the chapter. Click the **Add...** button:

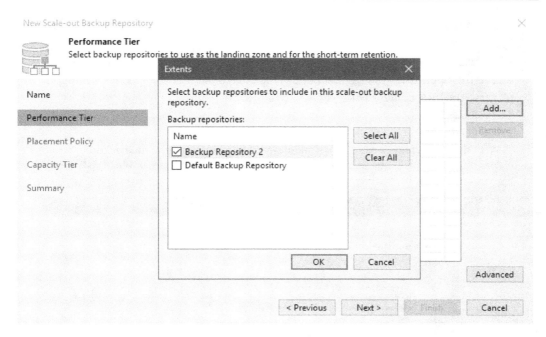

Figure 4.16 – Adding a performance tier from the standard repository

4. You then click **OK** and **Next**, which brings you to the **Placement Policy** section of the wizard. Here you choose between two options:

    a) **Data locality**: This option places all dependent backup files to the same extent, meaning all incremental and their full backup files. It can place subsequent backups to different extents as required. This option is typically selected when using Windows ReFS or Linux XFS to take advantage of the **Fast Clone** capabilities. It allows better deduplication and space savings within the repository.

b) **Performance**: This option allows the placement of full and incremental backup files to different extents. When used, it ensures better backup file transformation when using raw storage devices such as **raw device mappings (RDMs)**. This is due to the operating system having direct access to the storage and not having to traverse the virtual layer when using VMDK files:

> **Important note**
>
> With the **Performance** selection, should the extent containing the full backup files be lost, this will make restoration of the incremental files impossible.

Figure 4.17 – Data locality and Performance selection

5. After choosing **Placement Policy** and then selecting **Next**, you can select a **Capacity Tier** option for your scale-out repository. We will not cover this here, as we will go over this in one of the following sections of the chapter entitled *setting up an s3 object storage repository for use with a scale-out repository*.

6. You then click the **Apply** button, which creates the scale-out repository for you, and then click **Finish** at the **Summary** screen. You can now see the scale-out repository in the console view on the right-hand side:

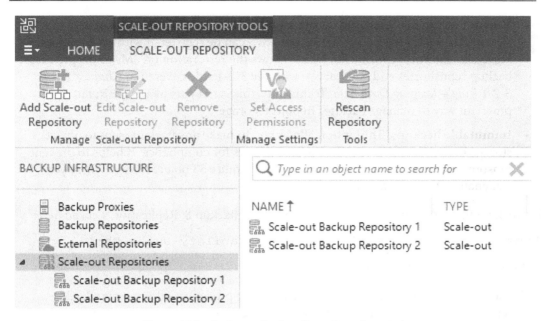

Figure 4.18 – Scale-out Backup Repository 2 created

We have now completed creating a standard repository and adding it to a scale-out repository. This allows us to have a backup target with at least one form of storage contained in it. Next, we will look at adding a standard repository for S3 object storage, which we will use for the capacity tier of the scale-out repository.

Setting up an S3 object storage repository for use with a scale-out repository

Object storage within Veeam Backup & Replication gets used for archiving data that requires longer-term retention and typically for long periods. Veeam Backup & Replication version 10 introduced some new enhancements to the object storage repository:

- **Backup import**: All backups in an object storage repository can now be imported with the click of a button, similar to how local file backups get imported. This process allows for easy restore should a disaster happen, allowing you to start restoring from backup copies that are within the object storage.

- **S3 operation performance improvements**: The rescan and retention processing now has multithreading, which enhances speed significantly.

- **Veeam Backup for Microsoft Azure and AWS support**: Backups created by Veeam Backup for Microsoft Azure and AWS can be registered with on-premises servers as external repositories, which allows the restoration of VMs to on-premises backup repositories and compliance with the 3-2-1 rule covered in *Chapter 2, The 3-2-1 Rule – Keeping Data Safe*. Technically, this is not part of an SOBR, but it is a powerful way to manage backups in object storage.

- **Immutable backups**: This option allows you to make any backups sent to object storage immutable for a specified period of days for compliance. It helps to prevent ransomware from deleting your files. It does require S3 object storage where you can enable *Object Lock* functionality.

For further enhancements, please refer to the Veeam Backup & Replication website here:

`https://www.veeam.com/whats-new-availability-suite.html`

Follow these steps to add an object storage repository:

1. Adding an object storage repository is the same process as adding a standard repository, except that you select the **Object storage** option in the **Add Backup Repository** window:

Add Backup Repository
Select the type of backup repository you want to add.   ✖

 Direct attached storage
Microsoft Windows or Linux server with internal or direct attached storage. This configuration enables data movers to run directly on the server, allowing for fastest performance.

 Network attached storage
Network share on a file server or a NAS device. When backing up to a remote share, we recommend that you select a gateway server located in the same site with the share.

 Deduplicating storage appliance
Dell EMC Data Domain, ExaGrid, HPE StoreOnce or Quantum DXi. If you are unable to meet the requirements of advanced integration via native appliance API, use the network attached storage option instead.

 Object storage
On-prem object storage system or a cloud object storage provider. Object storage can only be used as a Capacity Tier of scale-out backup repositories, backing up directly to object storage is not currently supported.

Cancel

Figure 4.19 – Add Backup Repository dialog – Object storage

2. After you click the **Object storage** option, you get prompted with the **Object Storage** dialog, where you select what type of object storage you want to add:

Figure 4.20 – Object Storage selection window

3. In this case, we will choose **S3 Compatible** to add a repository to Veeam Backup & Replication. Once you select this, a similar dialog to the standard repository appears, but is instead entitled **New Object Storage Repository**:

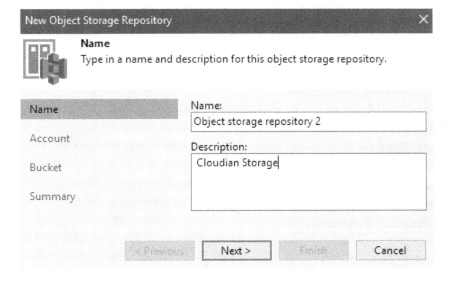

Figure 4.21 – New Object Storage Repository window

4. At this point, you need to type information in the **Name** and **Description** fields for the repository and then click **Next**. You will then be at the **Account** step, where you type in your account information for object storage:

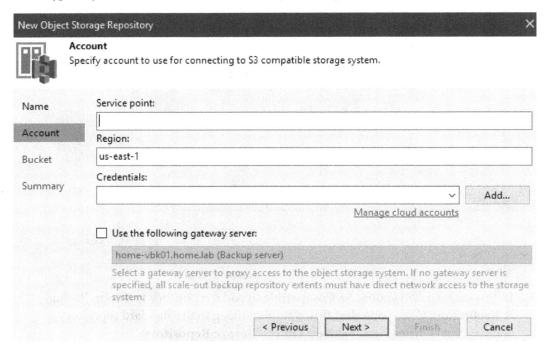

Figure 4.22 – Account information required for an object storage repository

5. At this point of the wizard, you enter the following information:

a) **Service point**: This is the URL that tells Veeam Backup & Replication how to connect to your object storage.

b) **Region**: This is typically a region name where the object storage is.

c) **Credentials**: These are the login credentials for the object storage, and if you have not added them to Veeam Backup & Replication, you can click the **Add...** button to do so.

d) **Use the following gateway server**: If you have set up a specific server to access the object storage repository, you will need to select the checkbox and choose the server from the drop-down list.

6.   Once you have entered all the information required, you click **Next**, which will validate the connection to the object storage. Then, the **Bucket** information comes up, which is usually already set on the object storage, but you can create a folder to place the backups in by clicking the **Browse...** button:

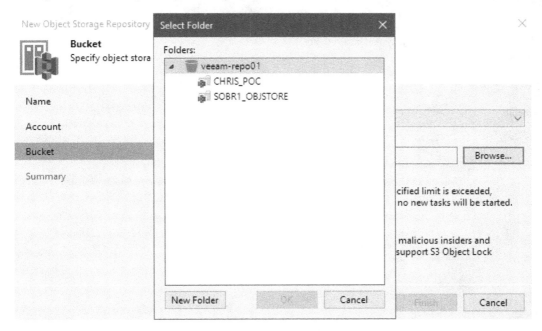

Figure 4.23 – Browsing to the Bucket folder or creating one

7.   If there was already a folder created for you, you can select it now or click the **New Folder** button to create a new folder for backups. After selecting or creating the folder, you click **OK**, which takes you back to the **Bucket** page.

8.   You will now have two options for your bucket:

a) **Limit object storage consumption to**: This allows you to set the specified storage amount allocated to you, or you leave it empty if there is no limit.

b) **Make recent backups immutable for**: This turns on an added layer of protection for *X* number of days. This setting protects against modification and deletion by ransomware, malicious insiders, accidental deletion, and hackers:

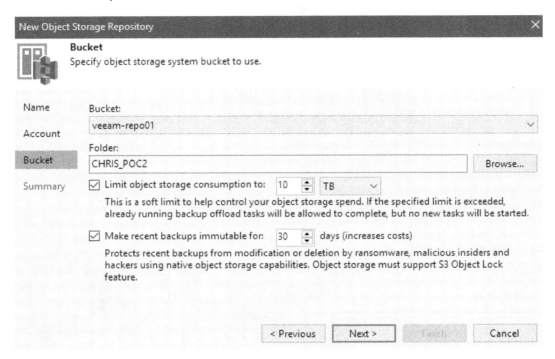

Figure 4.24 – Limiting storage and immutability options

> **Important note**
> Please also note that for the immutable feature, **S3 Object Lock** must be supported and turned on by the object storage vendor; otherwise, you will get an error when trying to proceed. Click on **Next** to continue to the **Summary** page and apply the settings.

We have now added an object storage repository to Veeam Backup & Replication. This will allow us to store our backups for longer-term retention when the offload task is run to move older backup points out to the capacity tier. We will now look at how this integrates into the scale-out repository in the next section.

Configuring the capacity tier for a scale-out repository, including the new copy policy

Now that you have added object storage as a standard repository, we can look at how you would integrate this into the scale-out repository as the capacity tier. We will also look at the new *Copy mode*, which is a part of this process:

1. To begin, you need to edit the scale-out repository we created in the steps outlined in the *Setting up a scale-out repository* section. Right-click the scale-out repository and select the **Properties** option to bring up the wizard. Click the **Next** button three times, which will take you to the **Capacity Tier** section of the **Edit Scale-out Backup Repository** window:

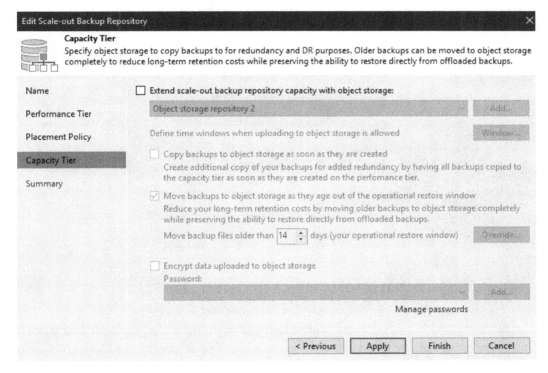

Figure 4.25 – Edit Scale-out Backup Repository and Capacity Tier option

2. At this part of the dialog, you need to place a checkmark in the **Extend scale-out backup repository capacity with object storage** option to enable it. You can now select the other options available:

    a) Repository to use: Choose from the drop-down list, or you can click the **Add...** button to go through the standard repository wizard as we did previously.

    b) **Window...**: This allows you to select the time when uploading to object storage can take place:

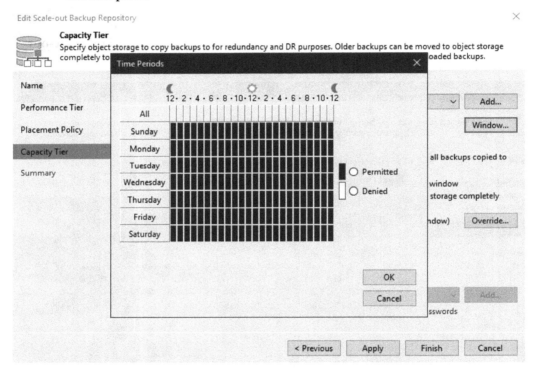

Figure 4.26 – Time window for uploading to object storage

Note that anything in colour is permitted, and anything white gets denied. You can select to upload overnight versus business hours if so desired.

c) **Copy backups to object storage as soon as they are created**: This will send the backups created in the performance extents as soon as they are available.

d) **Move backups to object storage as they age out of the operational restore window**: This helps reduce long-term retention costs by moving older backups to object storage by X number of days. Or, you can select to override the setting to move backups when the scale-out repository reaches a certain percentage:

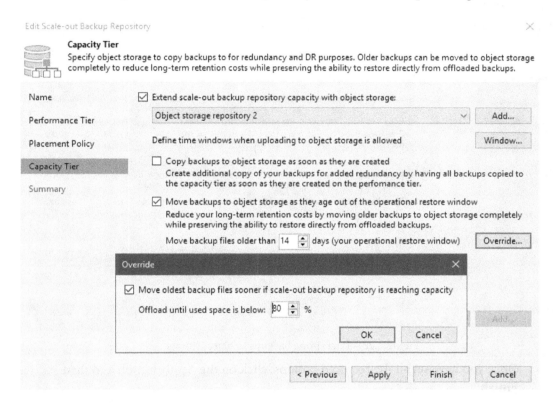

Figure 4.27 – Move backups by days or override by percentage

e) **Encrypt data uploaded to object storage**: This adds another layer of security to your backups in object storage along with the immutable feature:

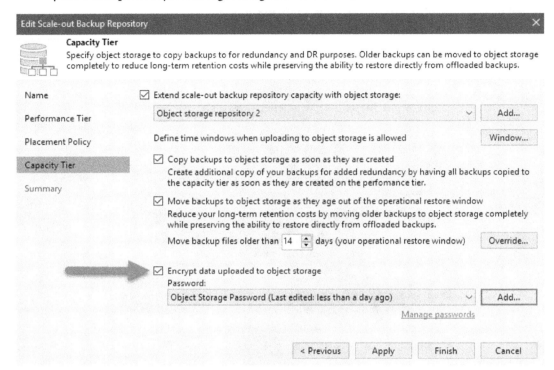

Figure 4.28 – Encrypting backups to object storage

3.   Once you have set all the required options, click on the **Apply** button and then **Finish**.

The preceding steps complete all of the requirements needed to set up a scale-out repository for your backups with additional settings for extra protection.

For more information about SOBRs, please visit Veeam's website at `https://helpcenter.veeam.com/docs/backup/vsphere/backup_repository_sobr.html?ver=100`.

Summary

This chapter has reviewed SOBRs and their components. We took a look at standard repositories and how users create them. We indicated the type of servers you can use for a repository in the scale-out, such as Windows (ReFS) and Linux (XFS). We covered how to add a standard repository to a scale-out. We also covered how to create a standard repository using object storage and how to set up the capacity tier of a scale-out to use this repository. After reading this chapter, you should now be able to work with scale-out repositories and be able to set up and configure them for use with backup jobs.

Hopefully, you will now have a better idea of scale-out repositories. The next chapter, *Chapter 5*, *Windows and Linux – Proxies and Repositories*, will take a deep dive into one of the standard repository features available for scale-out.

5
Windows and Linux – Proxies and Repositories

At the heart of Veeam Backup & Replication are the workhorses – *proxies* – and the central storage for backups – *repositories*. In this chapter, we will take a look at the proxy servers and what's new in version 10, including Linux proxies. We will discuss the differences between using Windows versus Linux as a proxy and some of the requirements. You will learn about the different proxy backup modes and use cases when each method is applied to your environment. We'll look at both Windows and Linux as repositories, as well as differences between the two. You will learn about the different filesystems that can be used for each repository and use cases of when to use each one. Finally, we will dive into job configuration and best practices for each of the filesystems for **Windows (NTFS/ReFS)** and **Linux (XFS)**.

By the end of this chapter, you will understand what proxy servers and repository servers are and what's new in version 10. You will also understand how each of them fits into your environment and use cases. Finally, you will have a better understanding of job settings and configuration best practices.

In this chapter, we're going to cover the following main topics:

- Understanding Windows and Linux proxy servers and what's new in version 10
- Distinguishing the differences between Windows and Linux proxy servers
- Configuring proxy servers and the different backup modes available
- Understanding Windows and Linux repository servers
- Discovering the different filesystems to be used with Windows and Linux
- Setting up jobs and configuration best practices with each filesystem

Technical requirements

For this chapter, you should have Veeam Backup & Replication installed. If you have been following along through the book, then *Chapter 1, Installation – Best Practices and Optimizations*, covered the installation and optimization of Veeam Backup & Replication, which you can leverage in this chapter. If you have access to both a Windows 2019 server and Ubuntu Linux 20.04, it will help with reinforcing the filesystem discussions.

Understanding Windows and Linux proxy servers and what's new in version 10

Within the Veeam Backup & Replication system, you have the backup server, which is the one that handles administering tasks, and the proxy servers sit between it and the rest of the backup infrastructure components. The proxy servers are the ones that process jobs and deliver the backup traffic to the rest of the infrastructure. You could say they are the *workhorse* of all the components.

Based on the following diagram, you can see that the proxy server performs many tasks before finally sending data to the backup repository:

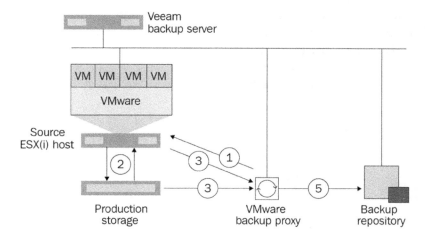

Figure 5.1 – Backup infrastructure with proxy server

The proxy servers will perform many tasks during a backup job, including the following:

- Retrieving VM data from within the production storage
- Compressing the data – during transit to the repository
- Deduplicating the data – deduplicates the data before sending it to the repository
- Encryption of the data – encrypts data in transit to the repository
- Sending of data to the backup repository (*backup job*) or possibly another proxy server (*replication job*)
- Performing restores

In a typical small deployment scenario, the proxy server is deployed on the backup server and is useful when you have low traffic loads. When it comes to larger Enterprise installations, a distributed structure and deploying proxy servers separately is better:

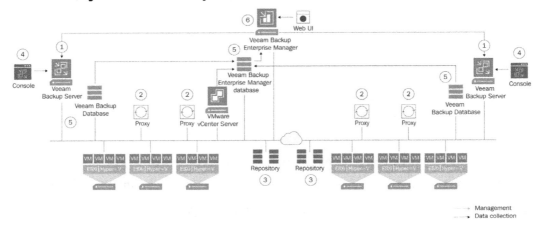

Figure 5.2 – Veeam components with a distributed architecture

This diagram shows an Enterprise environment infrastructure where the proxy servers are deployed separately on the network. Note that a proxy does not explicitly need to be a dedicated system, but for larger environments, this becomes a natural choice.

When you deploy a proxy server, two components get installed on the server:

- **Veeam Installer Service**: This is an auxiliary service installed and started on any Windows server once added to the managed servers within the Veeam Backup & Replication console. It analyzes the system and installs or upgrades components and services dependent on role selection.

- **Veeam Data Mover**: This is the heart of the proxy server and is the component that performs the data processing of tasks on behalf of the backup server and, as noted in the diagram, the proxy server.

With the release of Veeam Backup & Replication version 10, there were a couple of enhancements made to the proxy servers, mainly for Linux but also a few for Windows:

- **Linux backup proxy**: This is the version in which Linux proxies were introduced and are *hot-add proxies only* as they do not support any of the other transport modes. It provides the ability to deploy the user's choice of Linux distribution and offers greater security, knowing that proxies have direct and unrestricted access to production data. Future Veeam releases will expand Linux backup proxy capabilities.

- **Windows Server version 1909 support**: You can install backup infrastructure components on this operating system.

While this is not an extensive list of new things, using Linux for a proxy server is the most significant change for users. It allows the move away from Windows licensing, adding better security and using one of the many distributions, such as *Ubuntu, CentOS, RHEL, SLES*, and many others.

See the following Veeam Backup & Replication website for further information on the requirements of the proxy servers for both Windows and Linux:

```
https://helpcenter.veeam.com/docs/backup/vsphere/system_
requirements.html?ver=100#proxy
```

Now that you understand the proxy servers and what has changed in version 10, we will look at the differences between Windows and Linux proxy servers to better understand the right one for your environment.

Distinguishing the differences between Windows and Linux proxy servers

Deploying a proxy server works the same whether you select Windows or Linux, and they do the same job, which is task processing for the backup server. There are, however, some requirements and limitations for both Windows and Linux, which we will cover.

When you deploy a proxy server, the hardware requirements for both Windows and Linux are pretty much the same:

- **CPU**: An x86-64 processor with a minimum of two cores (vCPUs), plus one additional core per concurrent task

- **Memory**: 2 GB of RAM plus 200 MB for each concurrent task

- **Disk space**: 300 MB plus an additional 50 MB per concurrent task

- **Network**: 1 Gbps or faster for onsite backup and replication with 1 Mbps or faster for offsite backup and replication

---

Important Note

A task during the backup process on a proxy server is one *VMDK* file of a VM. So, if you have a backup job with five VMs, each with four disk (VMDK) files each, and your proxy server has eight vCPUs, it can process only eight of those VMDK files at one time of the 20 total for all VMs. The concurrent tasks topic was covered in *Chapter 1, Installation – Best Practices and Optimizations*.

Also, note on the Veeam Backup & Replication best practices site that when working in a virtual infrastructure, you should have proxy servers with a maximum of eight vCPUs and deploy more as required for task processing depending on the environment and load.

One limitation between both Windows and Linux proxy servers is using the **virtual appliance backup mode (HotAdd)** to process VM data. When you back up the proxy server itself with Veeam Backup & Replication, it will turn **Changed Block Tracking** (**CBT**) mode off to back up the proxy server due to other VMDK disks being attached. HotAdd mode is not allowed.

When it comes to the Linux proxy servers, there are a few more limitations to note:

- The account used must be either root or a user elevated to root.

- The disk.EnableUUID parameter of the Linux server must be set to true in the vSphere client.

- Only **virtual appliance transport mode (HotAdd)** is available for backups.

- You cannot use Linux backup proxies with VMware Cloud on AWS due to the VDDK settings required, which can't be enabled.

- You cannot use them for the following backup scenarios: replication over WAN accelerators, VM copy, file share backup, or integration with storage systems.

When it comes to the different **transport (backup)** modes, there is one standard mode between both Windows and Linux, which is the **virtual appliance backup mode**, also known as **HotAdd**. Many requirements must be met for the proxy server to use this mode:

- The proxy server role must be assigned to a VM – you cannot use a physical proxy server in this instance due to access to the hosts and storage. This holds true for backing up servers that reside on VMware vSAN as well.

- Disks of the backup proxy must reside on the vSAN datastore.

- vSAN backup is processed over the I/O stack of the ESXi hosts where the proxy server is deployed.

- Each of the proxy servers and VMs getting processed must reside in the same vCenter data center.

- The ESXi host on which the proxy server resides must have access to the datastore (storage) where the disks of the VMs you plan to process are.

- When VMs resides on an NFS 3.0 datastore and you have planned backups for them, you must put the proxy server on the same host where the VMs reside due to the potential for them to become unresponsive during snapshot removal, as noted by this VMware KB document: *Virtual machines residing on NFS storage become unresponsive during a snapshot removal operation* (2010953): `https://kb.vmware.com/s/article/2010953`.

- An alternative to NFS 3.0 is to use ESXi 6.0 and higher with NFS 4.1 to avoid this issue.

- Both the backup server and the proxy server must have the latest version of the VMware tools installed.

- SCSI 0:X must be present on the backup proxy server.

All of the preceding limitations, including others about the older VMFS3 datastore and block sizing, are listed on the Veeam Backup & Replication website here:

`https://helpcenter.veeam.com/docs/backup/vsphere/virtual_
appliance.html?ver=100`

Now that we have covered the proxy server differences between Windows and Linux, let's take a look now at configuration and transport mode selection when you deploy a proxy server.

Configuring proxy servers and the different backup modes available

When setting up Veeam Backup & Replication, as we have noted many times throughout the book, the proxy servers are among the more critical infrastructure components since they do all the task processing for running jobs. Therefore, setting them up and selecting the right **transport mode** for your backups is an essential step to ensure that you both achieve the performance and meet your backup window. Both job efficiency and the time required for job completion will significantly depend on the transport mode. All of the transport modes described apply to Windows servers, but for Linux, the only transport mode that it uses is the **virtual appliance** mode.

There are three modes that you can select, starting with the most efficient first:

- **Direct storage access**: In this mode, Veeam Backup & Replication reads/writes directly from/to the storage system where VM data or backups reside. There are two modes: **Direct SAN access** and **Direct NFS access**.

- **Virtual appliance access**: This mode is called **HotAdd**, and the disks of a VM are attached directly to the proxy server's SCSI controller to read/write data instead of using the network.

- **Network**: This mode is the least efficient; however, you can use it with any infrastructure configuration. Data is retrieved from the ESXi host over the network using the **Network Block Device** (**NBD**) protocol.

> **Important Note**
> When selecting network mode, depending on your proxy server setup, it can be just as good as, if not better than, the other transport modes when using 10 GB or higher networks – especially when using a physical proxy server. It is always a good idea, if your network permits, to have backups segregated on their own network.

The following chart shows the storage types and recommended transport modes to use:

Production Storage Type	Direct Storage Access	Virtual Appliance	Network Mode
Fiber Channel (FC) SAN	Install a backup proxy on a physical server with direct FC access to the SAN.	Install a backup proxy on a VM running on an ESXi host connected to the storage device.	This mode is *not recommended* on 1 GB Ethernet but works well with 10 GB Ethernet. Install a backup proxy on any machine on the storage network.
iSCSI SAN	Install a backup proxy on a physical or virtual machine.		
NFS Storage			
vSAN	Not supported.	Install a backup proxy on a VM running on an ESXi host connected to the vSAN storage device.	
VVol		Install a backup proxy on a VM running on an ESXi host connected to the vVol storage.	
Local Storage		Install a backup proxy on a VM on every ESXi host.	

Figure 5.3 – Storage and transport modes

As we touched on previously with the **Direct storage access** option, there were two modes noted that we will take a look at now:

- **Direct SAN access**: This mode should be used when you have data on shared *VMFS SAN* datastores connected to ESXi hosts. This can be via **Fiber Channel (FC)**, **Fiber Channel over Ethernet (FCoE)**, **Internet Small Computer Systems Interface (iSCSI)**, and **Shared SAS** storage. *VMFS* is the filesystem that VMware uses when creating a datastore to place all *VMDK* disk files.

- **Direct NFS access**: This mode should be used when your data resides on an NFS datastore. Veeam Backup & Replication uses the VMware VDDK to communicate with the ESXi host to retrieve the data; however, it puts an additional load on the host.

> **Important Note**
> Further information is available on the Veeam Backup & Replication website: *Transport Modes – Veeam Backup Guide for vSphere*:
> `https://helpcenter.veeam.com/docs/backup/vsphere/`
> `transport_modes.html?ver=100`

The following shows the process of configuring a Windows or Linux proxy server, including the transport mode selection within the Veeam Backup & Replication console:

1. Open the Veeam Backup & Replication console, click on the **BACKUP INFRASTRUCTURE** tab, and ensure you have **Backup Proxies** selected:

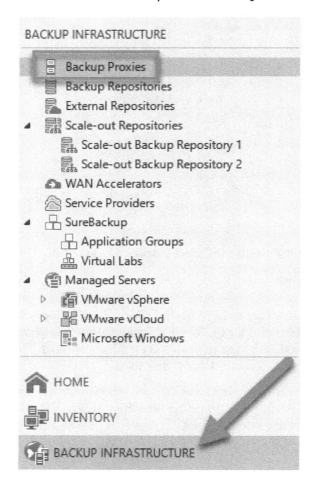

Figure 5.4 – BACKUP INFRASTRUCTURE – Backup Proxies

2. At this point, you can use the **Add Proxy** button in the toolbar or right-click in the window and select **Add VMware backup proxy…**, which will bring up the **New VMware Proxy** window:

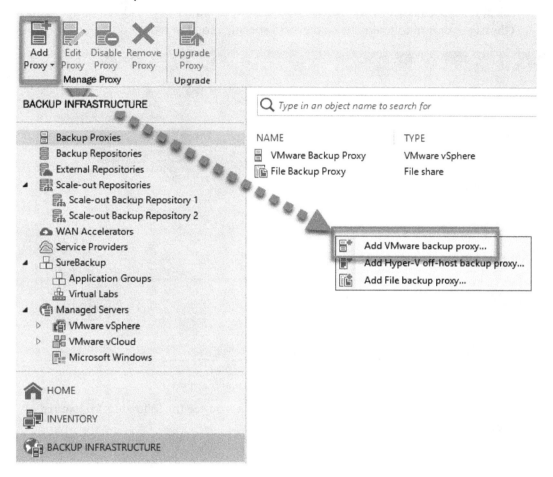

Figure 5.5 – The Add Proxy button or right-click options

3. You can now use the **Choose server** drop-down list to select either your Windows or Linux server to be used as a proxy server, and you can also click **Add New...** as well if your server is not showing. Also in this window is the **Transport mode** selection, which is set to **Automatic selection** by default, but you click on the **Choose...** button to make the change to a specific mode as required:

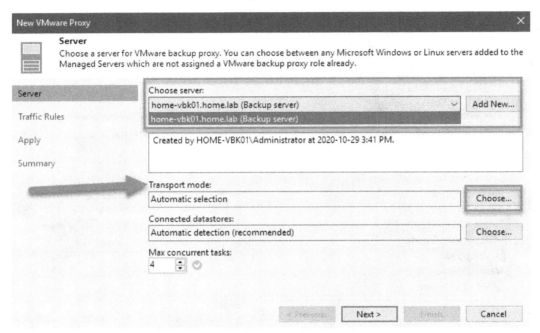

Figure 5.6 – Server selection for proxy and transport mode

4. When you click **Choose...** for **Transport mode**, you see the following **Transport Mode** dialog:

Figure 5.7 – Transport Mode dialog for the proxy server

This dialog is where you can either leave **Automatic selection** turned on, and Veeam Backup & Replication will choose the best method for backing up your servers, or you pick the transport mode that you wish to use. The **Failover to network mode if primary mode fails, or is unavailable** option is selected by default. Should you choose a transport mode that fails, then Veeam Backup & Replication will fall back to using network mode regardless of selection. Should you choose to use network transport mode, you also have the option to turn on **Enable host to proxy traffic encryption in Network mode (NBDSSL)**, which will ensure traffic gets backed up across the network using **SSL encryption**.

> **Important Note**
>
> For the **Transport mode** selection, when using a *Linux proxy server*, you can only use **Virtual appliance** or **Network** transport mode from the options presented due to the operating system's limitations.

Now that we have covered in detail the proxy servers and transport modes, we will now take a look at repository servers, both Windows and Linux, and the differences between each, including the filesystems that you will use within Veeam Backup & Replication.

Understanding Windows and Linux repository servers

When it comes to storing backups, the repository server is the primary target for backup files, VM copies, and metadata. You can use the following types of repositories:

- **Direct attached storage**: These can be virtual or physical and are typically Windows or Linux servers.

- **Network-Attached Storage (NAS)**: Network shares that are either SMB (CIFS) or NFS are used.

- **Deduplicating storage appliances**: These are hardware devices such as Dell EMC Data Domain, ExaGrid, HPE StoreOnce, and Quantum DXi.

- **Object storage**: These are cloud storage services you can use as backup repositories.

> **Important Note**
> Do not configure multiple backup repositories pointing to the same location or using the same path. However, this statement does not apply when using some newer multi-node-clustered NAS system shares. The configuration is different, especially when using a multi-node cluster within a scale-out repository.

When we discuss both Windows and Linux servers (virtual or physical), we are talking about the *Direct attached storage* noted previously. This type could be directly connected storage, FC, or even *iSCSI storage*. The repository drive gets connected to the server as either another VMDK file for a VM or another set of disks set up in a RAID group (DAS), via FC or iSCSI, typically for a physical server. When using a physical server for either Windows or Linux, there are some advantages as well as disadvantages.

Advantages of using a Windows physical server

- **Processing power**: You will have full use of the CPUs within the server where the virtualization layer is not a factor.

- **Memory**: You will have full use of the RAM in the system and not have memory sharing within the virtualization environment.

- **Disk processing**: Drives are connected directly to an SCSI card within the system and can be extremely fast at processing data, which can be advantageous.

Disadvantages of using a Windows physical server

- **Storage expansion**: Unless the server has an expansion card where you can connect via a **Serial-Attached SCSI (SAS)** cable, then you are limited to what is in the server.

- **Storage upgrade**: If you are unable to attach an expansion shelf of disks, then the only way to upgrade would be to migrate data off, upgrade the disks to larger sizes, then migrate the data back or start fresh.

- **Server recovery**: You would need to have something such as the Veeam Agent for Microsoft Windows installed to take backups and recover using bare-metal restore, which is processing the booting of the emergency CD created and pointing to the drive where the backups are stored.

Now, when it comes to virtual Windows or Linux servers, there are also advantages and disadvantages of both.

Advantages of using virtual servers

- **Scalability**: This can be on-demand and may not require additional hardware up to what the physical server hosting the VMs can support.

- **Management**: When it comes to maintenance, VMs are much easier to work with as failure of a physical server could take several days to restore, whereas with VMs, if located in a cluster of physical servers, they move between servers.

- **Portability**: When it comes to VMs, moving them across environments can be done quickly, whereas, with a physical server, it requires downtime for your backups as the repository would be offline during the move.

Disadvantages of using virtual servers

- **Licensing**: While you pay for both a physical server hardware when using a virtual server, the costs can be the same, if not more, depending on your environment.

- **Skillset**: Managing a virtual environment requires a higher level of skillset, which will require having IT members competent enough to handle the virtual infrastructure.

So, as you can see, there are both advantages and disadvantages to each type of environment; to further expand on this, the following chart shows a great comparison:

Physical Servers	Virtual Machines
Large upfront costs	Small upfront costs
No need for licensing purchase	VM software licenses
Physical servers and additional equipment take a lot of space	A single physical server can host multiple VMs, thus saving space
Has a short life-cycle	Supports legacy applications
No on-demand scalability	On-demand scalability
Hardware upgrades are difficult to implement and can lead to considerable downtime	Hardware upgrades are easier to implement; the workload can be migrated to a backup site for the repair period to minimize downtime
Difficult to move or copy	Easy to move or copy
Poor capacity optimization	Advanced capacity optimization is enabled by load balancing
Doesn't require any overhead layer	Some level of overhead is required for running VMs
Perfect for organizations running services and operations which require highly productive computing hardware for their implementation	Perfect for organizations running multiple operations or serving multiple users, which plan to extend their production environment in the future

Figure 5.8 – Physical versus virtual servers

Now that we have a good idea of physical and virtual servers for the repository, we can look at both the recommended filesystems used within a repository and what is best suited for use within a scale-out repository.

Discovering the different filesystems to be used with Windows and Linux

Once you have determined your choice of a physical or virtual server as your repository, you need to look at the filesystem when formatting the drives and what will work best for each scenario of backup. When it comes to both Windows and Linux, there are a few types of filesystems that work best for backups:

- **Windows NTFS**: This the original **NT FileSystem** (**NTFS**) and works well with any backup but is especially useful when backing up databases such as Microsoft Exchange. This is because Exchange databases are in JET format and take much better advantage of the 4K block sizing on NTFS.

- **Windows ReFS**: The newer **Resilient File System (ReFS)** is designed to optimize data availability, efficiently manage scalability, and ensure data integrity through **resilience** to file corruption.

- **Linux EXT4**: This is the **Linux fourth extended filesystem** and uses extents to replace traditional block mapping. It is built for journaling as well as speed versus reliability.

- **Linux XFS**: This is a high-performance 64-bit journaling filesystem. It excels in executing parallel I/O operations due to its design.

Of the filesystems we have listed, we will look further into the Windows ReFS and Linux XFS to show the advantages of using these for Veeam Backup & Replication repository servers.

The Windows ReFS was introduced in Windows 2012 and has improved in Windows 2016 and now Windows 2019. Using Windows 2019 with the ReFS filesystem allows you to take advantage of some great features:

- **Resilience**: This is a new feature introduced to accurately detect corruption and correct this corruption while remaining online, providing greater data integrity and availability. Integrity streams uses checksums for metadata and for file data, which allows file corruption detection.

- **Performance**: Improvements were made for virtualized and performance-sensitive workloads. There is real-time tier optimization, block cloning, and sparse **VDL** (short for **Valid Data Length**). You can also set variable cluster sizes using 4K or 64K block sizes; using 64K block size is very good for large sequential I/O workloads.

Of the listed features for ReFS, one of the main things to use with Veeam Backup
& Replication is **block cloning**. What block cloning helps with is accelerating copy
operations, which allows faster and lower impact for VM checkpoint merging. It also saves
space since it does not duplicate blocks when storing them. ReFS, when used as a standard
repository or as an extent in a scale-out repository, helps when using the incremental
backup feature with synthetic full operations. The synthetic full backup process takes
advantage of the block cloning feature, so it is good to use this in your backup jobs with a
Windows ReFS repository, as shown:

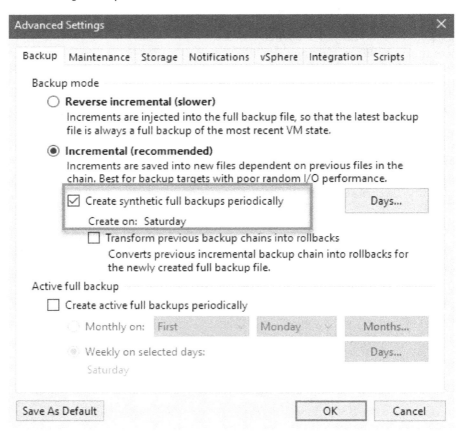

Figure 5.9 – Incremental backup with synthetic full

When you have this option selected in your backup job on the specified day, Veeam
Backup & Replication will create a synthetic full backup using the block cloning feature,
which takes much less time than if you were to do active full again. This process makes
your backups, in the end, much faster at completing.

> **Tip**
>
> When setting up your Windows 2019 server and formatting your backup data drive, ensure to select ReFS and the appropriate block size for your workload. If you will have many backups jobs with a lot of I/O to the repository, it is best to select the **64K** block sizing. Additionally, ensure the Windows Server 2019 system has the latest updates.

The *Linux XFS* filesystem has improved in the newer versions of Ubuntu 20.04 and 20.10. The recommended distribution of Linux to use for XFS is Ubuntu 18.04 (as per Veeam Backup & Replication; see, in the *Further reading* section, *Fast Clone in Veeam Backup & Replication*) or higher as they contain the changes made to take advantage of the fast cloning technology. Some of the newer features for XFS are as follows:

- **Capacity**: Since XFS is a 64-bit filesystem, it can scale very large and supports a maximum filesystem size of 8 exbibytes minus one byte ($2\wedge63 - 1$ byte).

- **Journaling**: Ensures consistency of data in the filesystem by writing the filesystem metadata to a serial journal before disk blocks are updated.

- **Online defragmentation**: A utility that allows the filesystem to be defragmented while it is mounted and online.

- **Online resizing**: Allows the expansion of the volume while mounted and online using the `xfs_growfs` command.

Fast Clone for Linux works on the `reflink` technology in the XFS filesystem. `reflink` allows the system to reduce disk space consumption and accelerate the copying of files. It will enable Veeam Backup & Replication to use the Fast Clone technology when using synthetic full backups, as noted in *Figure 5.9*. The following steps outline some of the processes to create the drive to be used as XFS within Veeam Backup & Replication:

1. To create a repository using the Linux XFS system, you must run the following command once you have the server configured and the hard disk to be the repository added:

   `lsblk` will list the block devices in Linux, which is where you will see the disk you added and used in the next command after `/dev/`:

   ```
   mkfs.xfs -b size=4096 -m reflink=1,crc=1 /dev/sda1
   ```

2. In this command, the last part, /dev/sda1, represents the disk inside the Linux server to be the repository drive. You must also create a folder and mount the folder to the newly formatted drive:

```
mkdir /backups
mount /dev/sda1 /backups
```

3. Once you have created the volume and mounted the folder, you will need to determine the UUID for the new drive as that will need to go into the /etc/fstab file, which will automatically load the drive on bootup:

```
blkid /dev/sda1
/dev/sda1:  UUID="fd4c0586-2d7b-4b11-87b4-721d8529875b"
TYPE="xfs"
```

4. You need to make a note of the UUID and then input that into the /etc/fstab file using the following command:

```
echo 'UUID=fd4c0586-2d7b-4b11-87b4-721d8529875b /backups
xfs defaults 1 1' >> /etc/fstab
```

You can now reboot your Linux server, and you will be ready to add it to the Veeam Backup & Replication console. Complete the following steps to add the Linux repository and ensure Fast Clone is used for backups:

1. From the Veeam Backup & Replication console, select the **BACKUP INFRASTRUCTURE** tab and click on **Backup Repositories** in the tree. Then, use either the **Add Repository** button in the toolbar or right-click to bring up the shortcut menu and select **Add backup repository…**:

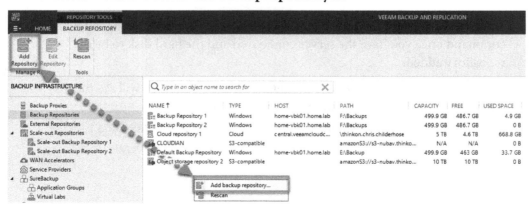

Figure 5.10 – Add Repository button or right-click menu

2. When the **Add Backup Repository** wizard comes up, select **Direct attached storage**:

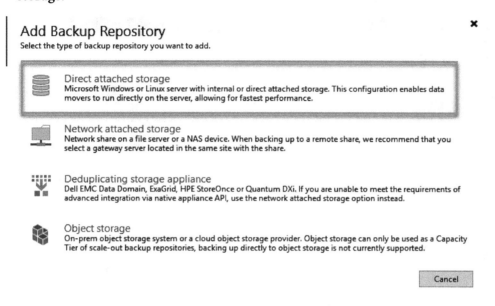

Figure 5.11 – Add Backup Repository wizard – the Direct attached storage option

3. Once you click **Direct attached storage**, select the **Linux** server option:

Figure 5.12 – The Linux server option

4. You will now see the **New Backup Repository** wizard, where you will need to create a name and description, then click **Next**:

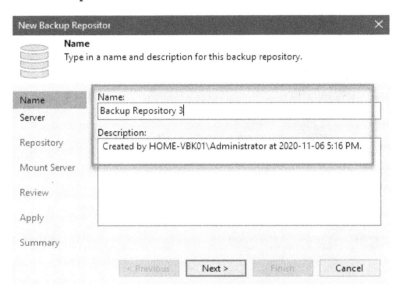

Figure 5.13 – Name and description of the Linux repository

5. On the next screen of the wizard, click on the **Add New...** button to add your Linux server to Veeam Backup & Replication for use as a repository:

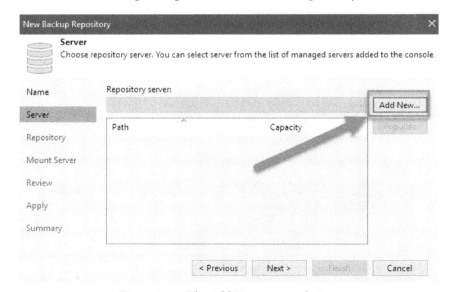

Figure 5.14 – The Add New... server button

6. The **New Linux Server** wizard will come up, and here you will need to type in the DNS name or IP address, add the SSH connection settings, trust the server and accept the SSL certificate, and then finish the wizard:

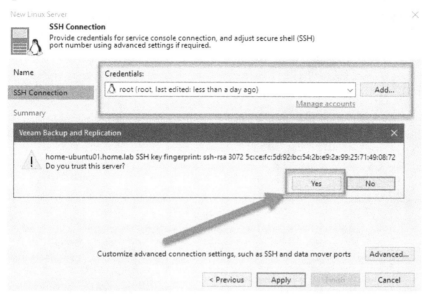

Figure 5.15 – New Linux server wizard credentials and SSH fingerprint acceptance

7. After completing this wizard, you are back at the **New Backup Repository** wizard, where you will need to click **Populate** to see the repository drive you created previously for XFS with the commands:

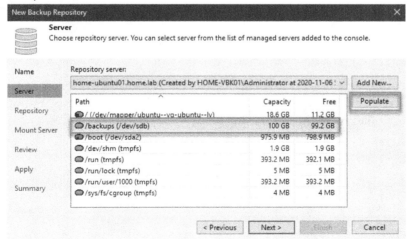

Figure 5.16 – Populating the XFS drive to be used

8. Click **Next** to proceed. The next screen shows the location selected and the
    important part for using Fast Clone, which is the **Use fast cloning on XFS volumes**
    checkbox, which you check:

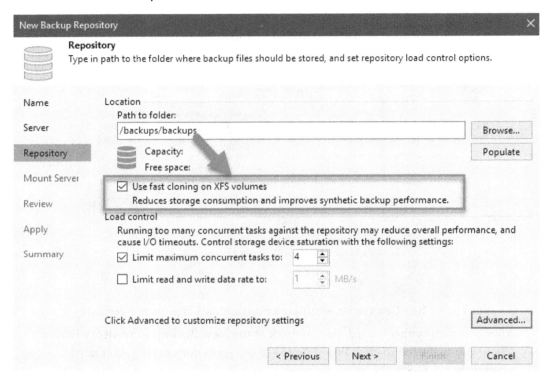

Figure 5.17 – The Fast Clone option for the XFS volume on Linux

9. You can then proceed through the remainder of the wizard, where you will select a
    mount server, review the settings, apply the settings, and finish the wizard.

These steps complete the process to set up XFS on Linux and add the server as a
repository to Veeam Backup & Replication. We will now look at job configuration best
practices for each filesystem in your backup jobs.

Setting up jobs and configuration best practices with each filesystem

The one thing to ensure when you are going to use Windows ReFS or Linux XFS for your repository server, whether it is a standard repository or part of a scale-out repository (extent), is that you are using incremental backups with the *synthetic full* option. It is the synthetic full option that will take advantage of the Fast Clone technology and allow backups to run much faster. It also will save space with the block clone technology within each of these filesystems, enabling you to store more data.

These settings are in the **New Backup Job** window under the **Storage** section. Click on the **Advanced** button, which will bring up the **Advanced Settings** dialog to select the **Incremental** option with synthetic full backups periodically:

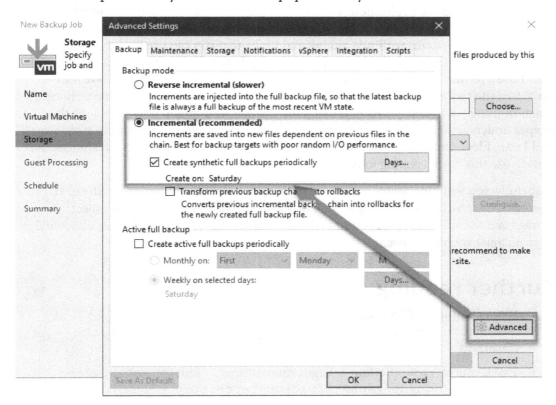

Figure 5.18 – Advanced job settings for use with Fast Clone on ReFS or XFS

When it comes to using the ReFS or XFS filesystems for your standard or scale-out repository, ensure that you keep the same filesystem and do not mix them. This holds true for the scale-out repository, and while the Veeam Backup & Replication documentation website states that you can mix them, it is best not to for better manageability.

You have now learned what option to set within your backup jobs to take advantage of the Fast Clone technology in the Windows ReFS and Linux XFS filesystems, as well as using the filesystem for repositories – standard and scale-out.

Summary

This chapter reviewed both Windows and Linux servers as proxies and repositories. We took a look at the differences in the proxy servers, as well as what was new within version 10 of Veeam Backup & Replication. We reviewed the different backup modes for the proxy servers and configuring them within Veeam Backup & Replication. We then took a look at the various filesystems within Windows and Linux that you can use for repositories, including both ReFS and XFS. We discussed how both Windows ReFS and Linux XFS could use a technology called **fast cloning** for fast backups and space-saving. We also discussed the setting within backup jobs that takes advantage of the Fast Clone technology when running your backups. After reading this chapter, you should now have a much deeper understanding of both proxy and repository servers regarding both the Windows and Linux filesystems. You should also now understand the ReFS and XFS advantages and how to use this when creating jobs.

Hopefully, you will now have a better idea of Windows ReFS and Linux XFS. The next chapter – *Chapter 6, Object Storage – Immutability*, will take a deep dive into object storage and the ability to make it immutable for preventing deletions (whether accidental or malicious) and ransomware.

Further reading

- Block cloning on Windows ReFS and how it works: `https://docs.microsoft.com/en-us/windows-server/storage/refs/block-cloning`

- Fast Clone in Veeam Backup & Replication: `https://helpcenter.veeam.com/docs/backup/vsphere/backup_repository_block_cloning.html?ver=100`

- Block cloning on Linux XFS and how it works: `https://blogs.oracle.com/linux/xfs-data-block-sharing-reflink`

- All about the XFS filesystem: `https://wiki.ubuntu.com/XFS?_ga=2.41569954.1261557641.1605031706-619614716.1605031706`
- vSAN backup considerations: `https://helpcenter.veeam.com/docs/backup/vsphere/virtual_appliance_mode_vsan.html?ver=100`

6
Object Storage – Immutability

One of the many options for backup storage and repositories is object storage, which now includes making backups stored here immutable. In this chapter, we will take a look at what Object Storage is and a brief history of it within Veeam Backup & Replication. We discuss how to use Object Storage within Veeam Backup & Replication. You will learn what Immutability is and how it applies to your environment. We will look at how Immutability helps to give you an extra layer of protection for your backups. Finally, we will dive into job configuration and best practices for using **Object Storage Immutability** features, including turning on immutable features.

By the end of this chapter, you will understand what Object Storage with Immutability is and when best to use it. You will also understand how this fits into your environment and use cases. And finally, you will have a better understanding of settings and configuration best practices.

In this chapter, we're going to cover the following main topics:

- Understanding Object Storage
- Explaining how to use Object Storage within Veeam Backup & Replication
- Understanding Object Storage Immutability and what it means

- Discovering how Immutability helps to protect your backup data
- Immutable backups – A great way to beat ransomware
- Working with Object Storage Immutability configuration and backup job settings

Technical requirements

For this chapter, you should have Veeam Backup & Replication installed. If you have followed along through the book, *Chapter 1, Installation – Best Practices and Optimizations*, covered the installation and optimization of Veeam Backup & Replication, which you can leverage in this chapter. If you have access to Object Storage with the Immutability options, that would be an added benefit, but is not required.

Understanding Object Storage

Object Storage is a computer data storage architecture that manages and stores information as **objects**, unlike filesystems, which use a file hierarchy and block storage that stores data as blocks within sectors and tracks. When data gets stored within object storage, it includes the data itself, an amount of metadata, and a **GUID** (short for **Globally Unique Identifier**). You can create a namespace within Object Storage, spanning multiple physical hardware instances, like a cluster.

You can compare Object Storage with other forms to see the differences:

- **Object Storage**: Takes pieces of data and designates them as an object, and then stores the data along with its associated metadata and a *GUID*.

- **File Storage**: Takes data and stores it in a hierarchy of folders to help organize it. This method is also known as hierarchical storage, which is similar to how paper files are stored, and data access is via a folder path.

- **Block Storage**: Data is broken down into singular blocks of data and then stored as separate chunks within the storage option. Each piece or block has a different address and, therefore, does not require a file structure.

Many will also refer to Object Storage as a **bucket** of storage, where you create a named bucket (namespace), and then when data gets sent to Object Storage, it is stored in the bucket. A great example of this would be with Amazon S3, as shown here:

Figure 6.1 – Amazon S3 object storage as a bucket

When it comes to Veeam Backup & Replication, Object Storage plays a role within the backup infrastructure in the repository section, including attaching the repository to a **SOBR** (short for **Scale-Out Backup Repository**), which is where it gets leveraged in the **Capacity Tier**.

> **Important note**
> Object Storage is not natively supported as a standalone repository to send backup jobs to directly, and you need to attach it to a SOBR as the Capacity Tier.

As noted below, you have everything go to the **Performance Tier** and then moved to **Capacity Tier**, which is where Object Storage gets used:

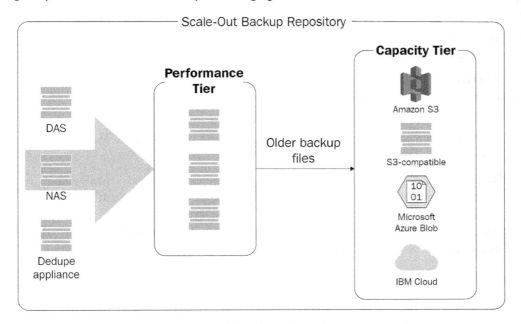

Figure 6.2 – Capacity Tier where Object Storage gets used

In the preceding diagram, you can see the different types of storage, with the most notable being the following:

- **Amazon S3**: This will include Standard, Glacier, and Glacier Deep Archive Storage.

- **Microsoft Azure Blob**: This is the Microsoft version of Object Storage.

- **IBM Cloud**: This is IBM Cloud Object Storage.

- **S3-compatible**: This can be cloud-based or on-premises with vendors, including Cloudian, Wasabi, and BackBlaze.

Of all the vendors listed, all of them are cloud-based, where they host the data, but if you use S3-compatible object storage, you can implement this on-premises.

Veeam Backup & Replication version 10 has made some significant enhancements in relation to Object Storage, as follows:

- **Backup Import**: All backups residing in an object storage repository can be imported with the click of a button, similar to how you import local backup files today. If you were to lose your infrastructure, but had backups that were in Object Storage, you could start to restore your backups immediately, including directly to the public cloud **IaaS** (short for **Infrastructure as a Service**) by using the free community edition of Veeam Backup & Replication.

- **Amazon S3 One Zone-IA Support**: This is a class of storage within Amazon for **Infrequent Accessed** (**IA**) backups, which can now get leveraged from Veeam Backup & Replication.

- **Microsoft Azure Data Box support**: There is native support available and a dedicated UI for registering Microsoft Azure Data Box storage solutions.

- **S3 operations performance improvements**: Both Amazon S3 and S3-compatible storage should see much faster rescan and retention processing thanks to the operations becoming multithreaded.

- **Veeam Backup for Microsoft Azure support**: You can now register Microsoft Azure blob containers created with Veeam Backup for Microsoft Azure as external repositories. This feature allows you to perform all types of restore and copy your Azure VM backups to on-premises backup repositories for disaster recovery and compliance with the 3-2-1 rule.

Now that we have taken a look at what Object Storage is and how it integrates with Veeam Backup & Replication, we will dive into the application to see how it is configured and used.

Explaining how to use Object Storage within Veeam Backup & Replication

To use Object Storage in Veeam Backup & Replication, you add it as a standard repository. You then use it in the Capacity Tier of the SOBR or as a target archive repository for *NAS Backup*. Veeam Backup & Replication cannot currently write directly to object storage; however, Veeam continues to enhance its object storage capabilities and have announced that more cloud tiers will be supported (such as archive tiers).

When using Object Storage with Veeam Backup & Replication, it is intended for long-term data storage and based on either a cloud solution or an S3-compatible storage solution that can be on-premises. When used as the Capacity Tier for a SOBR, it has a few useful features:

- **Storage space**: If your SOBR extents are starting to run out of space, adding Object Storage to the Capacity Tier allows you to tier off data to free up space in the Performance Tier.

- **Backup policies**: If your organization has policies or SLAs that require specific amounts of data, it will get stored on the extents and outdated or older data will be archived out.

- **3-2-1 rule**: Using the Capacity Tier with Object Storage in the cloud or other sites allows you to meet the 3-2-1 rule and make provision for disaster recovery more easily.

When using the Capacity Tier, there are some things that you can do with Object Storage:

- Move inactive backup chains to the capacity extent either by allowing the offload job to run, which is every 4 hours, or moving the backups manually:

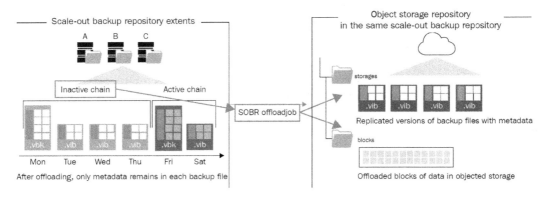

Figure 6.3 – SOBR offload process every 4 hours

The following screenshot shows how you can manually move the backups from Scale-Out to Capacity Tier – Object Storage:

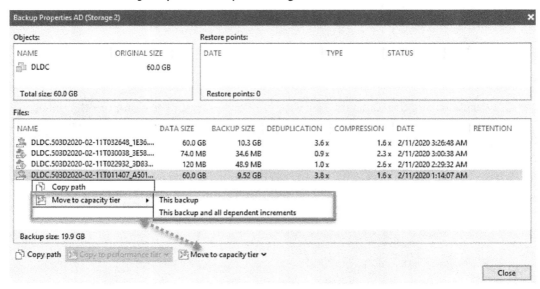

Figure 6.4 – Moving backups to the Capacity Tier – manual method

- Copy new backup files as soon as they are created in the **Performance Tier**:

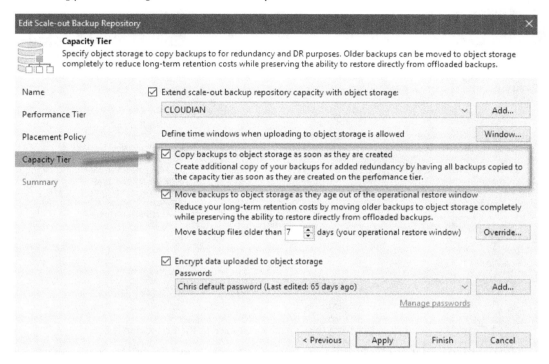

Figure 6.5 – Copying backups to Object Storage as soon as they get created

- Download data that was moved from the capacity extent back to the performance extents:

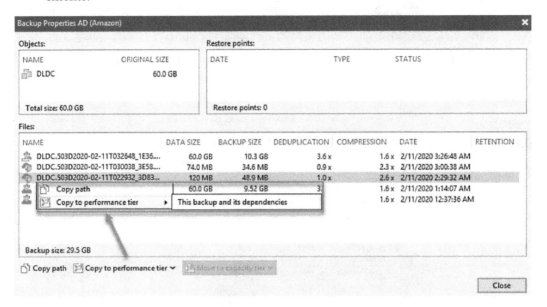

Figure 6.6 – Copying data from the Capacity Tier back to the Performance Tier

- Restoration of data directly from the Capacity Tier without having to create a new SOBR by importing backups once you make a standard repository pointing to your Object Storage:

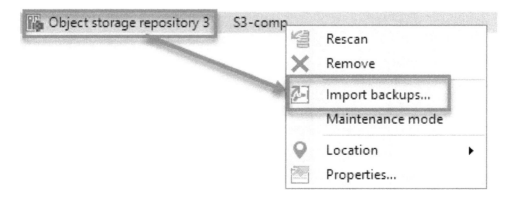

Figure 6.7 – Importing backups from the Object Storage repository for restoration

We have now covered the use of Object Storage within the Veeam Backup & Replication software, and will now dive into the Immutability of Object Storage in the next section.

Understanding Object Storage Immutability and what it means

When you decide to deploy Object Storage for use with Veeam Backup & Replication, an excellent option to look into is **Immutability**. When you turn this feature on within your Object Storage by selecting the **Object Lock** option, it prohibits objects from being changed or deleted throughout the storage lifetime or retention period. A storage lifetime is the time between successful object creation (upload) and successful object deletion.

The following diagram shows Object Lock and Versioning getting turned on:

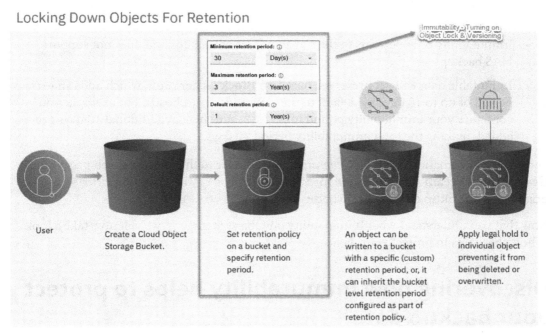

Figure 6.8 – Enabling Immutability, including the Object Lock feature

As you can see, turning on Object Lock and Versioning is a requirement for Veeam Backup & Replication to use this feature. Veeam Backup & Replication uses the *Object Lock* technology that Amazon and other S3-compatible providers have built in. Once this gets enabled in your Object Storage, it prevents the deletion of data from the *capacity extent* of your SOBR until the immutability expiration date passes.

When it comes to the Immutability feature for Veeam Backup & Replication, a number of limitations are as follows:

- When enabling Object Lock on the S3 bucket, ensure that the **None** option is selected for configuration mode. If not, you will not be able to register the bucket with Veeam Backup & Replication. Also, note that Veeam Backup & Replication uses **Compliance object lock** mode for each object it uploads.

- If you are using S3 Object Storage with backups created from Veeam Backup & Replication 9.5 Update 4, then you need to enable both *Versioning* and *Object Lock* at the same time before you turn on the *Immutability* feature. Any other approach will cause backup offload failures, and you will not be able to interact with the backups in the bucket.

- Immutability only pertains to the Capacity Tier of backups and does not support NAS backups.

- Immutable data will get preserved based on **Block Generation**, which adds an extra period of up to 10 days by default to the expiration date chosen. For example, you configure your immutability period to 30 days, and then an additional 10 days get added, making the total immutability period 40 days.

For further information on *Block Generation*, please refer to the following web page: **Block Generation – Veeam Backup Guide for vSphere**: `https://helpcenter.veeam.com/docs/backup/vsphere/block_gen.html?ver=100`.

Now that you understand what Immutability on Object Storage is, we will now take a look at how this helps to protect your data.

Discovering how Immutability helps to protect your backup data

If you decide to use the Immutability feature of Veeam Backup & Replication, you are taking steps to ensure that you are protecting your data. Turning this on protects your data from the following possible courses of action:

- **Deletion of data**: You cannot delete data on the Capacity Extent until the immutability expiration date passes, thereby preventing accidental deletion also.

- **Network attack**: If you have a breach in your network, having the Immutable feature prevents the attacker from compromising and deleting your data.

- **Malware activity**: Most malware or ransomware attacks try to encrypt or delete data, but having the Immutable setting prevents this and safeguards your data.

- **Other data-destructive actions**: If you have a disgruntled employee who was just relieved of their job, this will prevent any action they may attempt in terms of deleting or modifying the data.

Further to this list, Veeam Backup & Replication prevents the following things related to the program from being performed on immutable data:

- **Manual removal of data**: When trying to use the **Delete from disk** feature within the Veeam Backup & Replication console, if the data is on the Capacity extent, which is set to Immutable, then you cannot delete it.

- **Removal of data by retention policy**: When you have the number of retention points set in a job, the backups located in the Object Tier will not be deleted until the immutability period is over vis-à-vis the actual retention period.

- **Removal using cloud provider tools**: You cannot use any tools that a cloud provider might have to remove data that gets marked as Immutable.

- **Removal of data by cloud provider support**: Even the support department at a cloud provider is unable to remove or delete data from an Immutable object store.

- **Removal of data by the "Remove deleted items after" option**: Within the storage settings of a job under the **Maintenance** tab, there is a **Full backup file maintenance** section, and here you can turn on the **Remove deleted items data after** option to delete the data after a specified number of days, but this does not apply to Immutable backups:

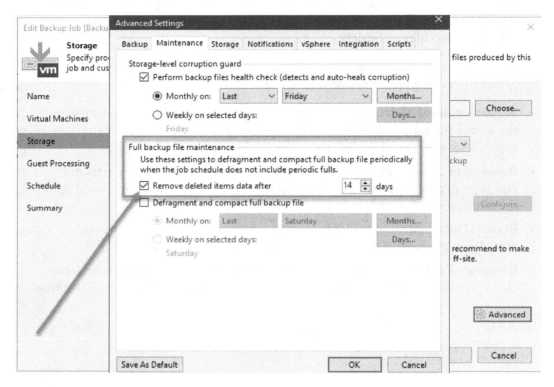

Figure 6.9 – The Remove deleted items data after option

Another way in which Immutability protects your data is where you use S3 Object Storage in the cloud, as it also takes into account the *3-2-1 rule* for backups. By having an offsite copy, including Immutability, you almost guarantee your data's safety unless there is a disaster at the cloud provider where your information is stored.

As you can see, Immutability protects your data in many ways, even within the Veeam Backup & Replication program. We will look at the settings you need to use within the application in the next section.

Immutable backups – A great way to beat ransomware

This is a special add-in section with guest content from *Rick Vanover, Senior Director of Product Strategy at Veeam*. You can interact with Rick on Twitter at `@RickVanover`:

This book by Chris is an excellent resource for those looking to do more with their implementation of Veeam Backup & Replication. When we consider some of the risks with data today, we need to take every advantage to be resilient against ransomware and other threats.

In my role on the Veeam product team, I have specialized in educating our customers and partners on what works to beat ransomware. I talk to customers often, and I enjoy hearing how customers can beat ransomware with Veeam. If you have experienced a ransomware incident, or know someone who has, there are several common themes. Some of these themes revolve around how the ransomware got in. Some stories revolve around how an organization struggled to determine what to do in terms of isolating this behaviour. Stories also relate to not knowing which parts of the infrastructure to trust. Stories will also indicate how the restore time or total time to resolution for this type of incident can be much longer than a typical outage.

A consistent theme when dealing with ransomware is ultra-reliance on a backup solution to get the organization out of this problem. What is also underscored is the fact that there is preparation to recover from this type of scenario. Sometimes, a backup is a saving grace, sometimes, storage snapshots are, on occasion, replication for disaster recovery is ready to go, as are a myriad of other combinations. What is most important, however, is the preparation that is required in order to recover from this type of scenario.

Preparation is important, and I'll return to that momentarily. When I see organizations not successfully remediate a ransomware situation, there is a common criterion. This common thread is a lack of preparation for this type of restore by not having one or more copies of data on some ultra-resilient media. An ultra-resilient media is some copy of data on a medium that is inherently resilient to this type of threat. I specifically recommend here at Veeam that organizations adhere to the 3-2-1 rule, with a specific focus on having at least one copy of data on an ultra-resilient media. Sometimes, this means a fourth copy of data, but this investment is well worth it when the ease of use of the cloud is considered against the risk of complete data loss.

An ultra-resilient media type can most easily be for backup data in the immutable backups of the Capacity Tier, as explained in this chapter. However, ultra-resilient media can also be for certain types of snapshots or offline storage. Specifically, an ultra-resilient media is something that has one, or more, of these characteristics:

- Offline

- Air-gapped

- Immutable

This will be the single most effective copy of backup data in terms of being resilient to ransomware threats. This also becomes a nice bonus line of defence against accidental deletion or certain malicious administrator-type activity. The following list of backup media types align to being ultra-resilient:

- **Tape media**: This is completely offline when removed from the library and not being read from or written to. Additional safeguards of **WORM** (short for **Write-Once Read Many**) media can be an additional benefit if a media is in a library.

- **Rotating hard drives**: This will have media removed from a computer intentionally and be offline. This approach is often associated with endpoint backups or small, remote sites. These systems will be completely offline when not in use.

- **Immutable backups**: The Veeam Scale-Out Backup Repository capacity tier supports immutable storage in the public AWS S3 cloud via service providers providing S3-compatible storage, storage with immutability and via on-premises unified storage systems providing S3-compatible storage with immutability. Each of these three scenarios is an immutable target for Veeam Backups.

- **Veeam Cloud Connect with Insider Protection**: I can't emphasize enough the added value of a service provider in general, and in the ransomware scenario, this is even more critical. Cloud Connect is Veeam Backup Storage as a Service. Insider Protection provides an out-of-band copy of the backup data that is resilient to tampering, accidental deletion, and so on and can be resilient to ransomware as well. The additional value of a service provider is the ability to provide a new target to restore to as well. This may be a helpful scenario in recovery.

- **Certain storage snapshots**: Some hardware arrays (primary or secondary storage) have snapshots or retention lock capabilities that are in place to help mitigate against a ransomware threat. If your storage system has a snapshot mechanism, ensure it is in use and research the specific capabilities when compared to ransomware threats.

Veeam has a future capability arriving in Veeam Backup & Replication v11, the Secure Linux Repository, which will allow backups to be placed on a Linux filesystem with immutable attributes. This is a versatile way to have *nearline Immutability* with no additional Veeam cost for backup data.

I've outlined the what, why, and how in terms of ransomware and Veeam, but what does this mean for you? This means it is time to ensure that the implementation you have is as resilient as it can be against these threats today and be prepared for the threats that come tomorrow. To help with this journey, Veeam has prepared several resources for customers and partners to wage war on the threats today for ransomware. Check out these resources:

- Ransomware assessment kit: `https://go.veeam.com/ransomware-prevention-kit`

- Ransomware education site: `https://www.veeam.com/ransomware-protection.html`

- Content library (executive and technical): `http://vee.am/ransomwareseriespapers`

We will now look at how you configure a repository with Immutability to use within the Veeam Backup & Replication console.

Working with Object Storage Immutability configuration and backup job settings

After you have your Object Storage set up with the Object Lock and Versioning features, you are then ready to configure the repository within the Veeam Backup & Replication console as follows:

1. Navigate to the **BACKUP INFRASTRUCTURE** tab and the **Backup Repositories** section of the console:

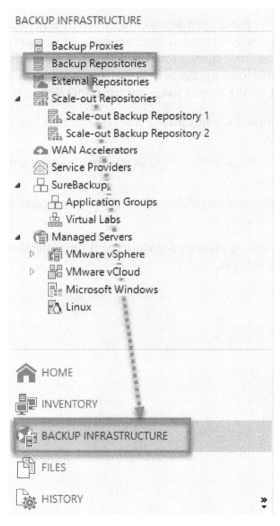

Figure 6.10 – BACKUP INFRASTRUCTURE tab and Backup Repositories

2. You can now click either the **Add Repository** button in the toolbar or right-click in the right-hand pane and select the **Add backup repository…** option:

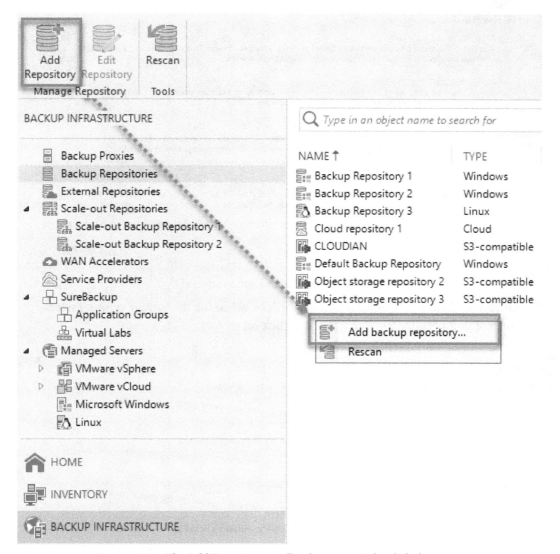

Figure 6.11 – The Add Repository toolbar button or right-click the menu

3. In the next window, you will select the **Object storage** option:

Add Backup Repository
Select the type of backup repository you want to add.

 Direct attached storage
Microsoft Windows or Linux server with internal or direct attached storage. This configuration enables data movers to run directly on the server, allowing for fastest performance.

 Network attached storage
Network share on a file server or a NAS device. When backing up to a remote share, we recommend that you select a gateway server located in the same site with the share.

 Deduplicating storage appliance
Dell EMC Data Domain, ExaGrid, HPE StoreOnce or Quantum DXi. If you are unable to meet the requirements of advanced integration via native appliance API, use the network attached storage option instead.

 Object storage
On-prem object storage system or a cloud object storage provider. Object storage can only be used as a Capacity Tier of scale-out backup repositories, backing up directly to object storage is not currently supported.

Cancel

Figure 6.12 – Object storage option selected

4. Once you click on **Object storage**, you will be asked for the storage type, so select the one that pertains to your setup and click on it:

 ## Object Storage
Select the type of object storage you want to use as a backup repository.

 S3 Compatible
Adds an on-premises object storage system or a cloud object storage provider.

 Amazon S3
Adds S3 object storage or an AWS Snowball Edge appliance.

 Microsoft Azure Blob Storage
Adds Microsoft Azure blob storage. Both Azure Blob Storage and Azure Data Box are supported.

 IBM Cloud Object Storage
Adds IBM Cloud object storage. S3 compatible versions of both on-premises and IBM Cloud storage offerings are supported.

Cancel

Figure 6.13 – Object Storage type selection

5. After selecting your Object Storage type, you will be prompted for a name and description, so fill in the **Name** and **Description** fields and then click **Next**:

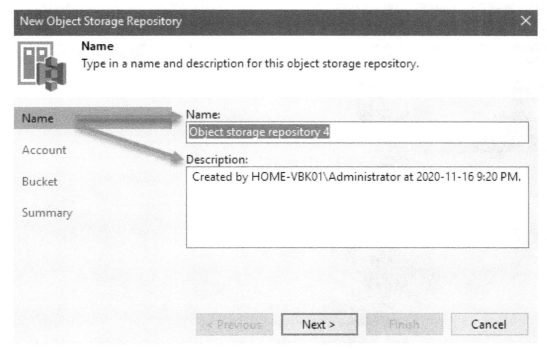

Figure 6.14 – New Object Storage Repository dialog – Name and Description

6. At the next screen, you need to type in the required information for your Object Storage – **Service point**, **Region**, **Credentials**, and **Use the following gateway server** if required. You may need to click the **Add...** button to enter the necessary credentials if they are not already available in the drop-down list:

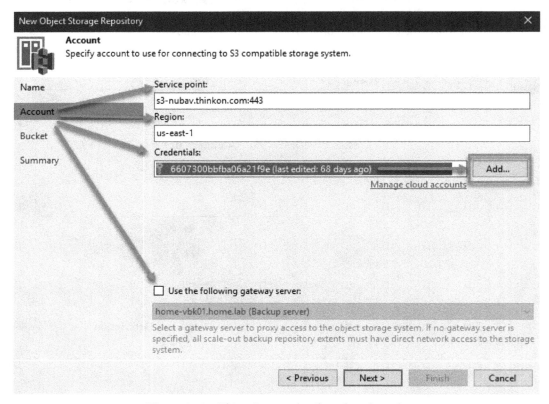

Figure 6.15 – Object Storage details and credentials

7. After you click **Next**, you enter the bucket information, which is where you will turn on the *Immutability* feature by selecting the checkbox next to **Make recent backups immutable for:**

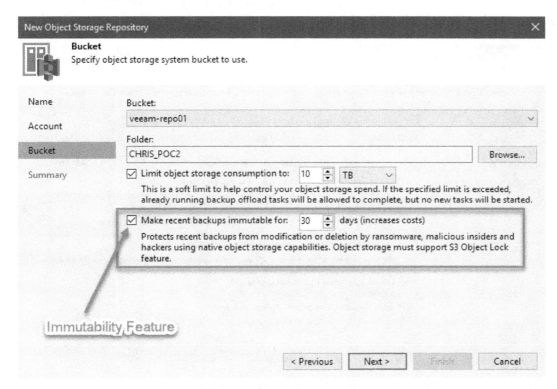

Figure 6.16 – Immutability feature enabled

8. To finalize the setup, click **Next** and then **Finish** to close the add repository wizard.

That completes the addition of an Object Storage repository, including the Immutability feature enabled. To use this, you add the repository as a *Capacity Tier* extent in your SOBR, which will then inherit the Immutable feature:

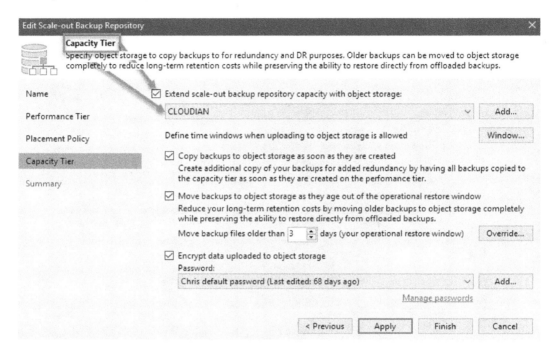

Figure 6.17 – Capacity Tier of Scale-Out to use the Immutability feature

You have now learned how to set up a standard repository for Object Storage and enable the Immutability feature, which then gets used in the Capacity Tier of the SOBR.

Summary

This chapter has reviewed Object Storage and provided a brief explanation of what it is. We took a look at how to use Object Storage within version 10 of Veeam Backup & Replication. We reviewed what Object Storage Immutability is and why it is essential and added a storyline of context in terms of being resilient to ransomware. We then took a look at when Immutability can protect your data and how. We also discussed the setting within Veeam Backup & Replication when creating a standard Object Storage repository and how to turn on Immutability. After reading this chapter, you should now have a much deeper understanding of both Object Storage and Immutability. You should also now understand how to configure the Immutable option within Veeam Backup & Replication and how it applies when using Scale-Out backup repositories and the Capacity Tier.

Hopefully, you will now have a better idea of Object Storage Immutability. The next chapter, *Veeam Datalabs*, will take a deep dive into On-Demand Sandbox, Secure Restore, Application Groups, and SureBackup Jobs.

Further reading

- Capacity Tier: Capacity Tier – Veeam Backup Guide for vSphere: `https://helpcenter.veeam.com/docs/backup/vsphere/capacity_tier.html?ver=100`

- Moving backups to the Capacity Tier (Object Storage) offload process: `https://helpcenter.veeam.com/docs/backup/vsphere/capacity_tier_move.html?ver=100`

- Manually moving backups to the Capacity Tier (Object Storage): `https://helpcenter.veeam.com/docs/backup/vsphere/moving_to_capacity_tier.html?ver=100`

- Downloading from the Capacity Tier: `https://helpcenter.veeam.com/docs/backup/vsphere/capacity_tier_download.html?ver=100`

- Immutability in Veeam Backup & Replication: `https://helpcenter.veeam.com/docs/backup/vsphere/immutability.html?ver=100`

Section 3: DataLabs, Cloud Backup, and Veeam ONE

The objective of this section is to teach you about the new Linux proxy option as well as Windows proxy options. We will dive into what makes up the Veeam DataLab components and how to use them for a variety of things, such as DEV and QA. We will look at service providers for offsite backups using the Veeam Cloud Connect option. You will gain a better understanding of which proxy best fits your environment and when to use each one. You will be able to set up a DataLab with all the components for testing. You will also learn what RansomGuard is and why it is an excellent option to choose for offsite backups. You will take a glimpse in to Veeam ONE and the monitoring capabilities that it has as well as reporting.

This section contains the following chapters:

- *Chapter 7, Veeam DataLabs*
- *Chapter 8, Cloud Backup and Recovery Using Veeam Cloud Connect Provider and the Insider Protection Feature*
- *Chapter 9, Instant VM Recovery*
- *Chapter 10, Veeam ONE*

7
Veeam DataLabs

Veeam Backup & Replication has significant components built into the software that can create a **DataLab** that allows you to do a variety of testing as well as DEV/QA work. In this chapter, we will take a look at the Veeam DataLab and what it is explicitly. We discuss the different components that make up a Veeam DataLab. We look at how you configure all of the required components to operate a Veeam DataLab. Finally, we will dive into how you go about using the Veeam DataLab for a variety of things, such as troubleshooting a VM, testing software patches and upgrades, and installing software.

By the end of this chapter, you will be able to explain what a Veeam DataLab is and the components required. You will define each part's requirements and know how to set up and configure them within the software. And finally, you will have a better understanding of when to best use a Veeam DataLab.

In this chapter, we're going to cover the following main topics:

- Understanding Veeam DataLabs – What are they?
- Distinguishing between the components that make up a Veeam DataLab
- Configuring the Veeam DataLab components within Veeam
- Understanding how to best use a Veeam DataLab

Technical requirements

For this chapter, you should have Veeam Backup & Replication installed. If you have followed along through the book, then *Chapter 1, Installation – Best Practices and Optimizations*, covered the installation and optimization of Veeam Backup & Replication, which you can leverage in this chapter.

Understanding Veeam DataLabs – What are they?

Veeam DataLabs is an included component of Veeam Backup & Replication, which enables availability and security by reducing malware risks, deploying upgrades and patches, allowing DevOps/Test, and allowing for the mitigation and remediation of compliance risks.

Some of the many useful features you can implement using the Veeam DataLab include the following:

- Automated backup and replica verification using **SureBackup** and **SureReplica**.

- **OnDemand Sandbox** allows you to bring up servers in a segregated environment and enable users to access this without affecting production.

- **OnDemand Sandbox** from storage snapshots allows you to leverage data and workloads from application-consistent fast storage snapshots and, in other cases, secondary systems.

- The ability to restore backups without the fear of malware or ransomware using the **Secure Restore** feature. This allows you to use an antivirus scanner during the restore process.

- The ability, during the restore process, to remove data while the restore operation is taking place using **Staged Restore**.

- **DevOps/Test** allows for the installation of patches or application updates in a controlled environment to ensure it works as expected before moving to production.

In order for you to use Veeam DataLabs, three components are required:

- **Virtual lab**: This is a small Linux appliance that runs within your environment and provides a gateway to your "lab" environment, allowing nothing to pass back into your production environment.

- **Application Group**: This is a concept where many workloads do not work alone, and they require multiple instances to test the functionality of the overall application. The application group will group the components and dependencies.
- **SureBackup job**: This is the policy-based schedule and group of when and where the sandbox environment is to run. This job brings together the *Virtual Lab* and *Application Group*.

Now that we have an overall understanding of what Veeam DataLabs are, their uses, and what they comprise, we will now take a more in-depth look into each component. Keep in mind that a Veeam DataLab is the common name (or a marketing name) of the combination of a Virtual Lab, Application Group, and a SureBackup job.

Distinguishing between the components that make up a Veeam DataLab

Veeam DataLabs can also be referred to as an **On-Demand Sandbox**. The three components that make up the On-Demand Sandbox can be created via the user interface or PowerShell. This process requires the following elements:

- Virtual Lab
- Application Group
- SureBackup job

Once these three components get configured, you can begin to take advantage of Veeam DataLabs, as it is a one-time setup.

We have taken a brief look at the components that make up the Veeam DataLab. We will now examine the parts of the Veeam DataLab to get a better understanding of what each one does.

Virtual Lab

Virtual Lab is an isolated environment that Veeam Backup & Replication uses to verify VMs. Veeam Backup & Replication starts VMs from the application group and the verified VM. The virtual lab is not only used for SureBackup verification, but also for **U-AIR** (short for **Universal Application-Item Recovery**), *On-Demand Sandbox*, and *staged restores*.

> **Important note**
>
> **Universal Application-Item Recovery** is an option in Veeam Backup &
> Replication that enables you to restore individual objects from virtual
> applications, such as email messages, database records, and directory objects
> from Active Directory. It leverages the vPower technology within the virtual
> lab.

The virtual lab does not require additional resources; however, VMs running in the virtual
lab consume the CPU and memory resources of the ESXi host where the virtual lab gets
deployed. Any changes made to VMs in the virtual lab get written to redo log files stored
on the datastore selected in the virtual lab settings and removed once the recovery process
completes.

The networking for the virtual lab is a fully fenced off environment from your production
network. The network configuration of the virtual lab mirrors the production network.
For example, if verified VMs and VMs from the application group are located in two
logical networks in production, the virtual lab will also have two networks. The virtual
lab networks get mapped to the same networks as production. As seen in the following
diagram, the isolated network has the same IP addresses as the production network:

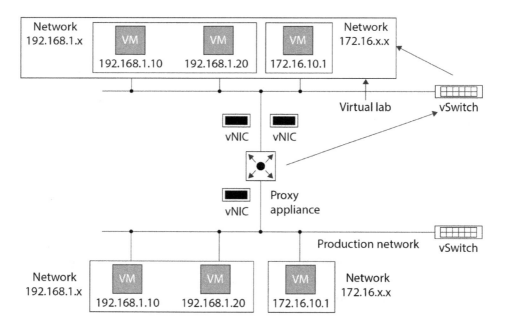

Figure 7.1 – Virtual lab segmented and isolated network

We have covered Virtual Lab and will now look at Application Group next.

Application Group

Most VMs work with other services and components in a network, like an email server with Active Directory. To verify such VMs, you must start all services and components on which the VM is dependent. Here, the *Application Group* is used within Veeam Backup & Replication.

The application group contains one or several VMs on which the verified VM is dependent. It will run applications or services that must start to enable the verified VM in order to function fully. Usually, you would have at least a **domain controller, DNS server,** and **DHCP server** in an application group. When the application group gets configured, you specify the roles of each VM, its boot priority, and boot delay. These settings allow for services to come up in the correct order before testing your verified VM. You are also able to specify which tests get performed to verify the VMs in the application group.

When you launch a SureBackup job, Veeam Backup & Replication starts VMs from the application group in the virtual lab in the required order and performs tests against them. Once the VMs from the application group get started and tested, the verified VM gets started in the virtual lab.

> **Important note**
> All VMs added to the application group must belong to the same platform –
> VMware or Hyper-V. Mixed application groups are not supported.

SureBackup job

A *SureBackup job* is a task for recovery verification. The SureBackup job will aggregate all settings and policies for the verification recovery task, such as the application group and virtual lab to be used, and VM backups that must get verified in the virtual lab. When a SureBackup job runs, it will create the environment for recovery verification:

1. Veeam Backup & Replication starts the virtual lab.

2. Once the virtual lab runs, the VMs from the application group get started in the required order. The VMs in the application group will remain running until the verified VMs (VMs from a linked job) are booted from backups and tested. If there is no valid restore point found for any VMs from the application group, the SureBackup job will fail.

3. Once the virtual lab is ready, Veeam Backup & Replication will power up the verified VMs (VMs from the linked job) to the necessary restore point. Now, depending on the job settings, the verification process begins one by one or can create several streams to verify several VMs simultaneously. If there is no valid restore point found for the verified VMs, the verification process fails for that particular VM, but the job continues to run.

By default, you can start and test up to three VMs at the same time. You also have the option to increase this number, but keep in mind that if any of the VMs getting verified are resource-intensive, then the performance of the SureBackup job, as well as the ESXi host, may decrease.

After the verification process is completed, the VMs in the application group are powered off. Still, you can choose to leave them running to perform manual testing or enable user-directed, application item-level recovery. In some rare cases, the SureBackup job may overlap the backup job that you have linked. If the backup job is running, it locks the files, preventing the SureBackup job from completing its verification. In this case, the SureBackup job starts once the primary backup job ends. You can chain the backup and SureBackup jobs and define a timeout period for the SureBackup job.

Please see the following diagram depicting the SureBackup job process:

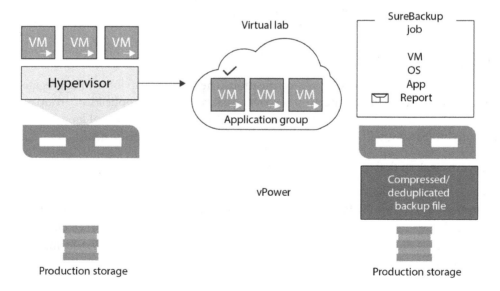

Figure 7.2 – SureBackup job process

Along with the main components in the preceding list for a standard On-Demand Sandbox, there is an option for the *On-Demand Sandbox* to use *Storage Snapshots* to allow you to power on VMs for testing, troubleshooting, training, and so on. The following storage systems allow the On-Demand Sandbox functionality:

- Dell EMC Unity XT/Unity, VNX(e)
- HPE 3PAR StoreServ, including secondary volumes – HPE 3PAR Peer Persistence
- HPE StoreVirtual P4000 series and HPE StoreVirtual **VSA** (short for **Virtual Storage Appliance**)
- IBM Spectrum Virtualize, including secondary IBM volumes – IBM Spectrum Virtualize HyperSwap
- NetApp, including secondary arrays – NetApp SnapMirror and NetApp SnapVault
- HPE Nimble storage, including secondary arrays – Nimble Snapshot Replicated Copy
- Universal Storage API Integrated Systems

The configuration of the On-Demand Sandbox for storage snapshots gets set up with the same components present in the regular On-Demand Sandbox:

- Virtual Lab
- Application Group
- SureBackup job

To start a VM from a storage snapshot in the On-Demand Sandbox, Veeam Backup & Replication presents the storage snapshot to an ESXi host as a datastore. The following actions get completed by Veeam Backup & Replication:

1. Veeam Backup & Replication detects the latest storage snapshot for the VM disks located on the storage system.
2. Veeam Backup & Replication triggers the storage system to create a copy of the storage snapshot. The snapshot copy helps protect the primary storage snapshot from changes.
3. The snapshot copy gets presented as a new datastore to the ESXi host on which the Virtual Lab is registered.
4. Veeam Backup & Replication performs regular operations required for the On-Demand Sandbox – reconfigures the VMX file, starts the VM, performs the necessary tests, and so on.

5. Once you finish working with the VMs and power off the On-Demand Sandbox, the cleanup operations take place – power off the VM and proxy appliance in the virtual lab, unmount the datastore from the ESXi host, and trigger the storage system to remove the snapshot copy.

The following diagram depicts these steps:

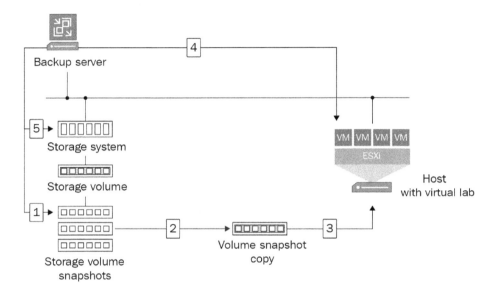

Figure 7.3 – On-Demand Sandbox with Storage Snapshot integration

You will now have a better understanding of each of the components that make up the Veeam DataLab. We will go through the configuration process for each element to show the settings and options available.

Configuring the Veeam DataLab components within Veeam

Now that you have a better understanding of each component of the Veeam DataLab, let's consider how to set them up within the Veeam Backup & Replication console. You will need to follow these steps next.

Configuring the virtual lab

Let's follow these steps to configure the virtual lab:

1. First, you will need to open the Veeam Backup & Replication console and go to the **Backup Infrastructure** section to select **SureBackup**:

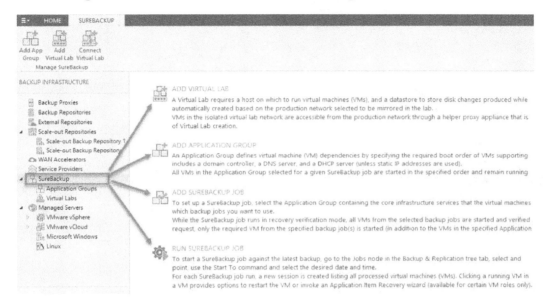

Figure 7.4 – SureBackup section of the Veeam console

2. The first thing to create is the virtual lab, so select the **ADD VIRTUAL LAB** option toward the right-hand side of the screen:

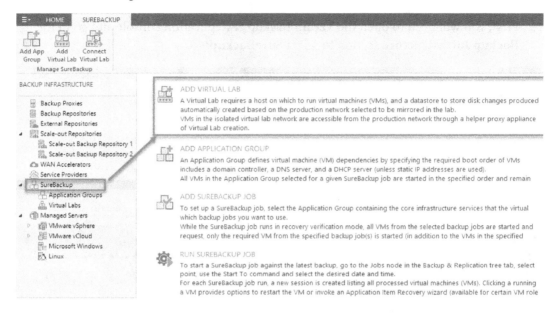

Figure 7.5 – The ADD VIRTUAL LAB option

3. At the first screen of the **New Virtual Lab** wizard, you specify a name and description:

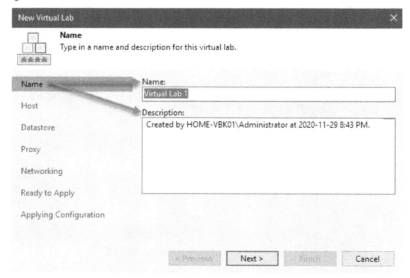

Figure 7.6 – New Virtual Lab wizard – Name

4. Click **Next**, and then you are brought to the **Host** screen to select which ESXi host to run the virtual lab:

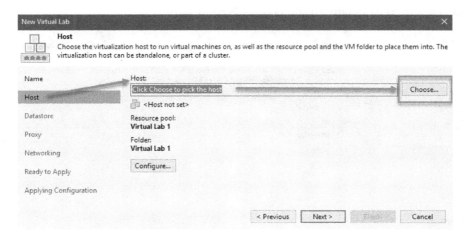

Figure 7.7 – Virtual Lab – Host selection

5. Click the **Choose** button and select your ESXi host. Once set, click **OK** to return to the **Host** screen of the wizard, where you can click the **Configure...** button to specify the resource pool and folder names:

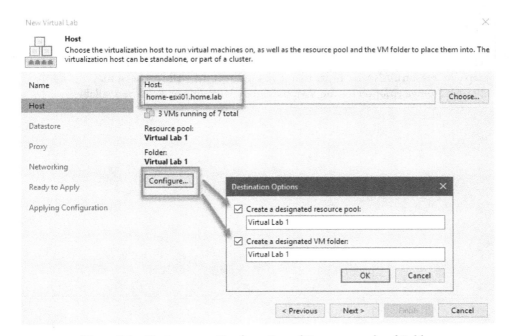

Figure 7.8 – Host screen – Configuration of Resource pool and Folder

6. Once the **Host** screen is complete, click on **Next** to proceed to the **Datastore** screen of the wizard. This screen is where you can redirect the changed blocks' location, which is stored by default in the vPower NFS location:

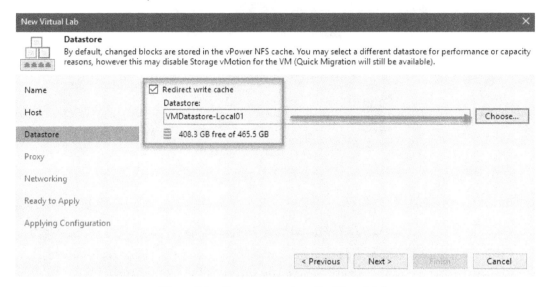

Figure 7.9 – Datastore options for Virtual Lab

Important note

As seen on the **Datastore** screen, you may want to change the location for changed blocks (*write cache*) to a different datastore for performance or capacity reasons. Also, be aware that if you redirect the write cache, it may disable *Storage vMotion* for the VM, but *Quick Migration* will still be available.

---

7. Once ready, click on the **Next** button to advance to the **Proxy** section of the wizard. Here you specify the Virtual Proxy Appliance settings and also determine if the proxy appliance will act as an internet proxy for the VMs in the lab:

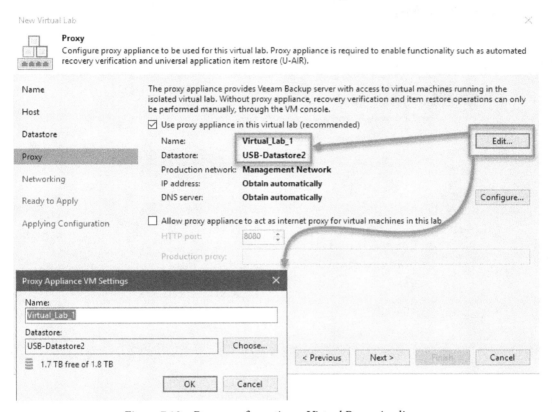

Figure 7.10 – Proxy configuration – Virtual Proxy Appliance

8. You can also configure the network settings by clicking on the **Configure** button to specify the **Production network, IP address**, and **DNS servers**:

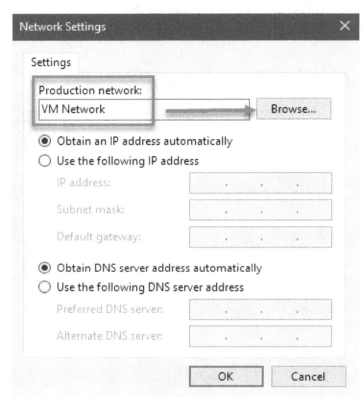

Figure 7.11 – Configuration of network settings for the virtual lab

9. Once you have the settings configured, including the correct network for your production VMs, click on the **Next** button to go to the **Networking** screen of the wizard. On this screen, you can select the networking option that best suits your environment:

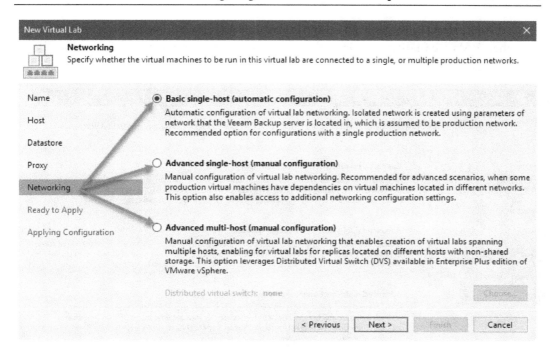

Figure 7.12 – Networking selection of wizard

The different options here are as follows:

a) **Basic single-host**: This is the most straightforward configuration option as Veeam Backup & Replication will set up the isolated network based on where the Veeam server is, which typically is your production network. It is recommended for a single production network.

b) **Advanced single-host**: This option requires manual configuration and gets used when VMs are in different networks. It allows for added network configuration settings:

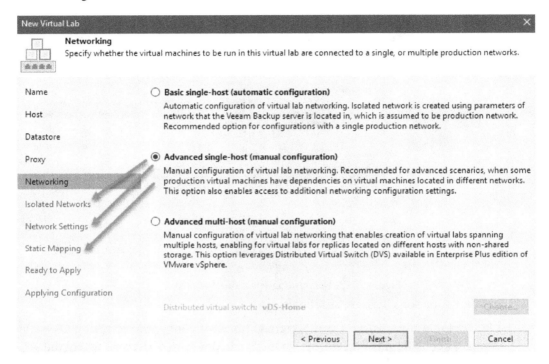

Figure 7.13 – Advanced single-host configuration option

c) **Advanced multi-host**: This option also requires manual configuration and allows access to multiple hosts instead of a single host. It also provides access to **Distributed Virtual Switch (DVS)**, available in the Enterprise Plus edition of VMware vSphere:

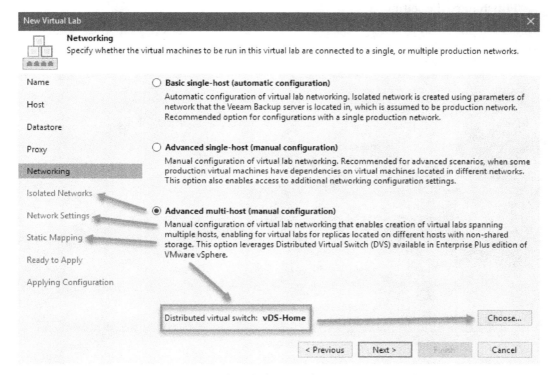

Figure 7.14 – Advanced multi-host configuration option with DVS

10. Once you have selected the corresponding option for networking (in the example, we will use **Basic single-host**), you click on the **Next** button to continue to the validation screen. Had you selected **Advanced single-host** or **Advanced multi-host**, then you go through the network settings screens noted below. You then click **Finish** once the settings are applied:

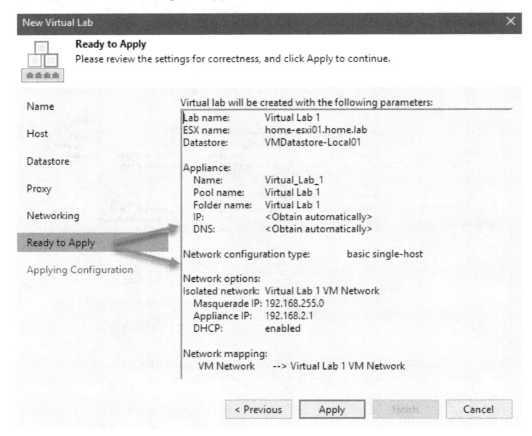

Figure 7.15 – Basic single-host configuration verification

Had you selected either **Advanced single-host** or **Advanced multi-host**, you would have needed to go through the networking screens of the wizard, shown as follows:

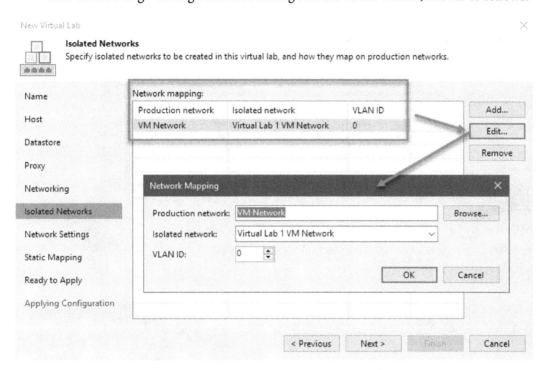

Figure 7.16 – Advanced single-host – Isolated Networks configuration

The preceding showed the **Isolated Networks** settings, and here is the next screen – **Network Settings**:

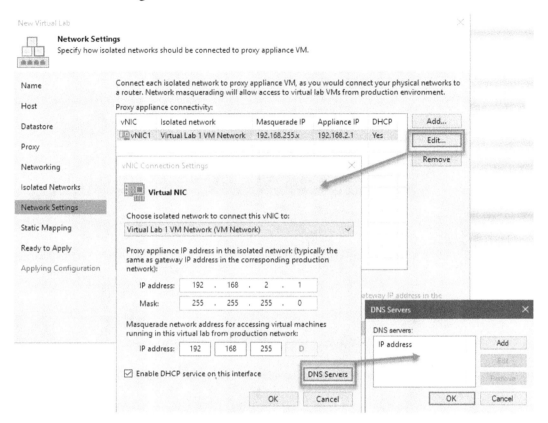

Figure 7.17 – Network Settings – Isolated network

> **Tip**
> Network segmentation uses **Network Masquerading** to give an IP address that cannot be routed to the production network to keep it in isolation.

11. The last screen for the networking component is **Static Mapping**:

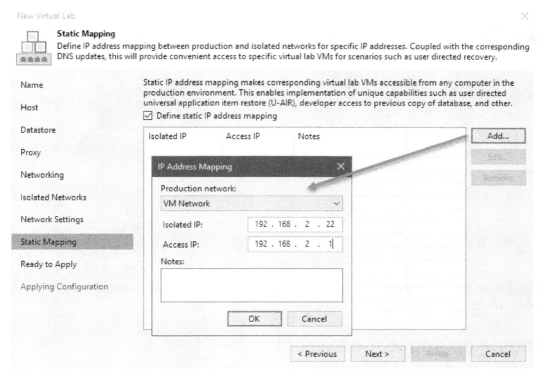

Figure 7.18 – Static IP mapping for accessibility

After you complete the *Virtual Lab* wizard, you can navigate to the **Virtual Labs** section under **SureBackup** to see your lab displayed and ready for use:

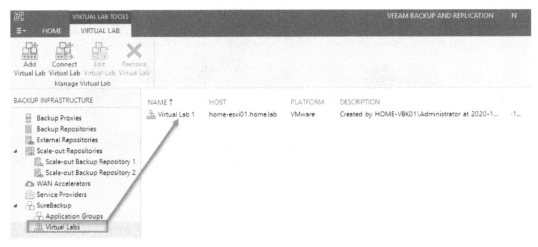

Figure 7.19 – Virtual lab created and ready

Now that we have completed setting up the *Virtual Lab*, we can look at setting up the *Application Group* required for testing VMs.

Adding an application group

Let's follow these steps to add an application group:

1. Click on **SureBackup**, as shown in the following screenshot:

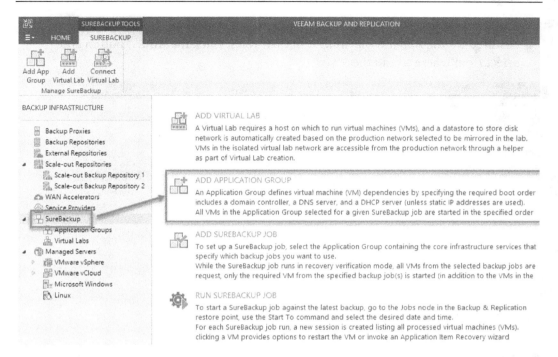

Figure 7.20 – The ADD APPLICATION GROUP option

2.   Click on the **ADD APPLICATION GROUP** option to start the wizard. Enter the name and description and then click **Next**:

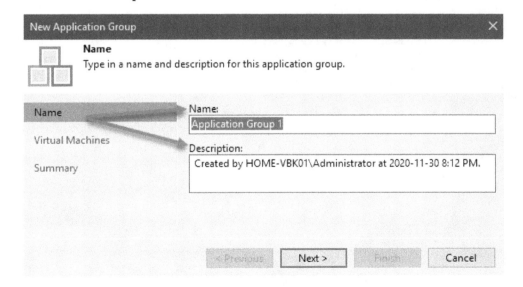

Figure 7.21 – The New Application Group wizard

3. You will now be at the **Virtual Machines** section, where you choose the VMs required for the *Application Group* and their roles. Click the **Add...** button and select from where to add the VMs:

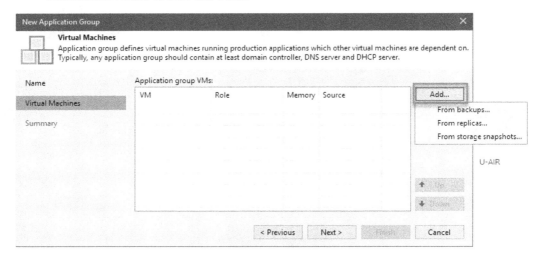

Figure 7.22 – Add Application group VMs

4. After clicking on the **Add** button and selecting the source, you choose the VMs:

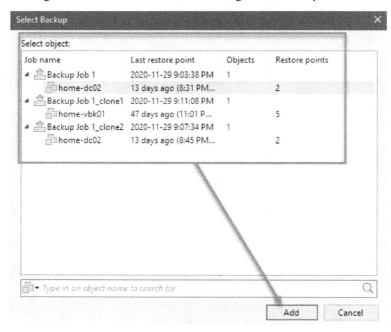

Figure 7.23 – Add VMs to Application Group

5. Once you have added the VMs, you then assign the required roles to allow for proper verification with the SureBackup job:

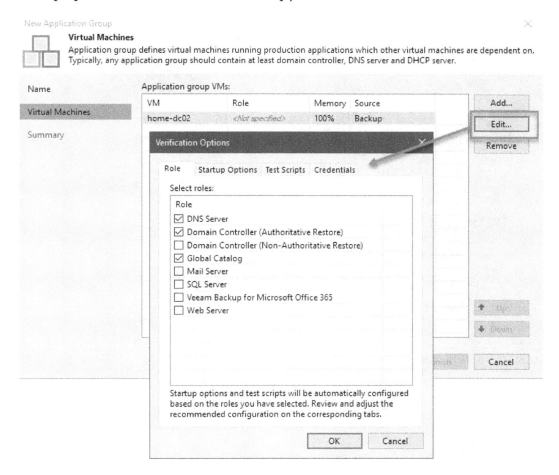

Figure 7.24 – VM roles in the application group

You will notice that when selecting a domain controller, there is an option to use either **Authoritative Restore** or **Non-Authoritative Restore**. Please refer to the *Further reading* section at the end of this chapter to explain these types of restores.

There are also other options that you can configure in the **Verification Options** dialog:

a) **Startup Options** – This includes elements such as memory allocation, startup time, and boot verification options – VM heartbeat and ping response:

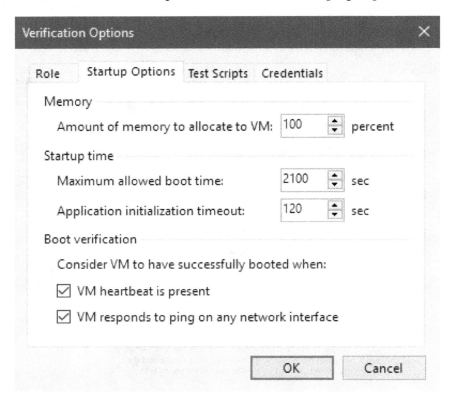

Figure 7.25 – Startup Options

b) **Test Scripts** – Depending on the roles selected, the **Test Scripts** tab gets populated with scripts already available in Veeam Backup & Replication, which you can edit, or you can add your own using the **Add** button:

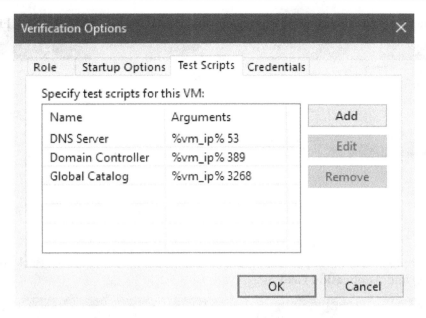

Figure 7.26 – Test Scripts

c) **Credentials** – These are the credentials used to complete all tasks within the VMs selected:

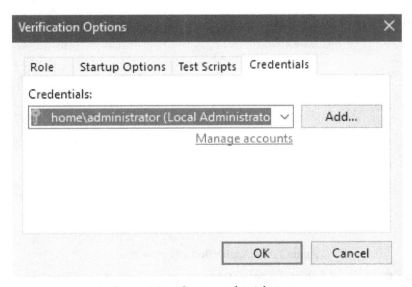

Figure 7.27 – Server credentials to use

6. The final screen is **Summary**, where you validate the settings and then click **Finish**:

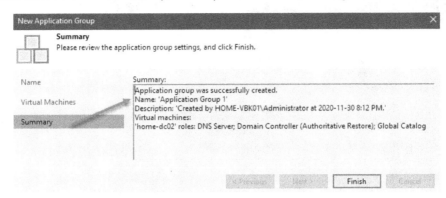

Figure 7.28 – Application group summary

You can now see your application group in the console under the **SureBackup** section and **Application Groups**:

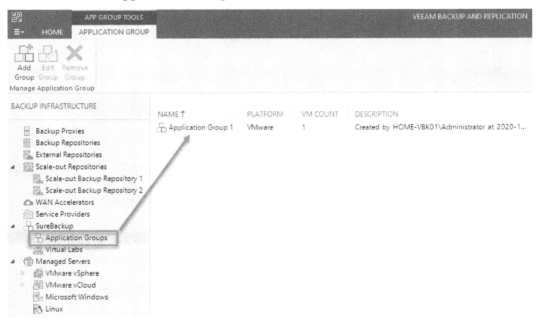

Figure 7.29 – Application group created

Once you have the *Virtual Lab* and *Application Group* configured, you can create a *SureBackup job* to automate validation of your backups.

Creating a SureBackup job

Let's follow these steps to create a SureBackup job:

1. Click on **SureBackup**, as shown in the following screenshot:

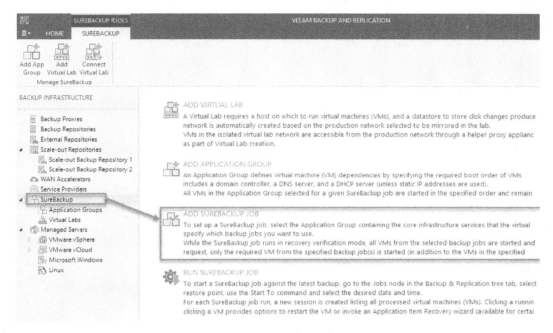

Figure 7.30 – SureBackup job option

2. Click on the **Add SureBackup Job** option to start the wizard, where you will specify a **Name** and **Description**:

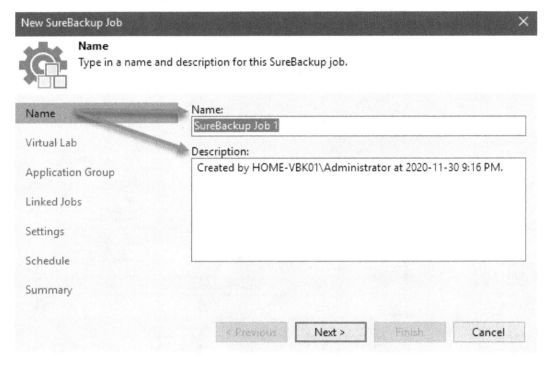

Figure 7.31 – SureBackup job wizard

3. Click **Next** to proceed to the **Virtual Lab** selection dialog, where you can now choose the lab you created in the preceding step:

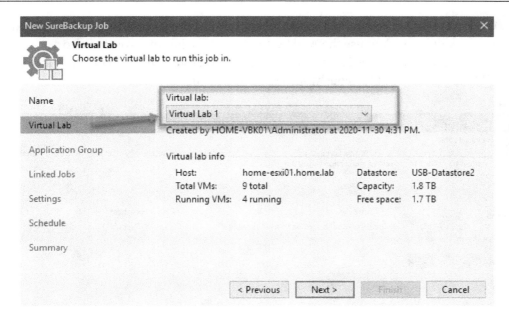

Figure 7.32 – Virtual Lab selection

4. Click **Next** to then select the **Application Group** to use created previously. There is also an option to allow the *Application Group* to keep running after your *SureBackup job* completes in case of further manual validation:

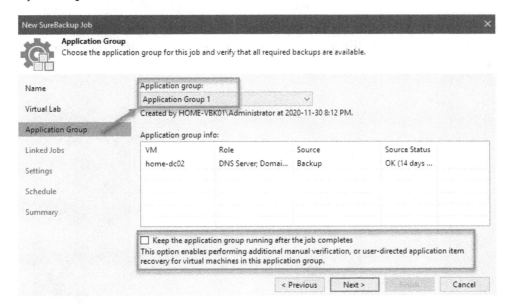

Figure 7.33 – Application group selection

5. Click **Next** to proceed to the **Linked Jobs** screen, where you will select the required job for validating. You use the **Add** button to add the job, and you can also set advanced verification tasks using the **Advanced** button. You also set the number of VMs to process simultaneously, with the default being three:

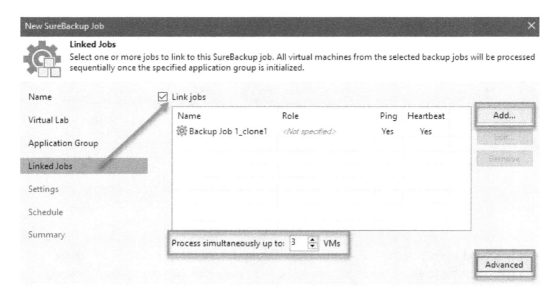

Figure 7.34 – Linked Jobs selection for validation

6. Click **Next** to proceed to the **Settings** screen. This screen is where you select things such as **Backup file integrity scan**, **Malware scan** (Secure Restore), and set **Notifications**:

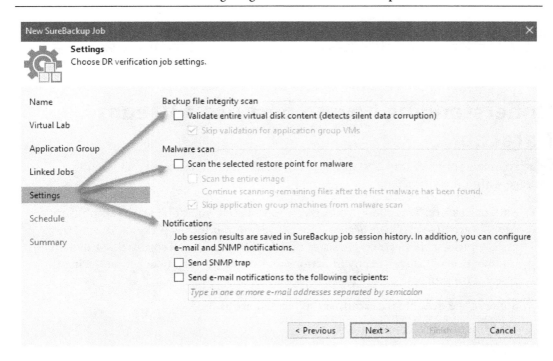

Figure 7.35 – Settings selection screen

7.   Click **Next** to proceed to set the schedule and review the summary. Click **Apply** and then **Finish** to complete the setup:

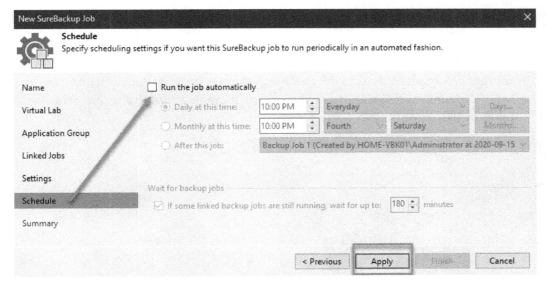

Figure 7.36 – Schedule screen and applying settings

This process now completes an entire Veeam DataLab from the *Virtual Lab*, *Application Group*, and *SureBackup job*. We will now look at use cases involving the use of a Veeam DataLab.

Understanding how to best use a Veeam DataLab

Once you have your Veeam DataLab set up, there are several use cases that you can leverage within the virtual lab:

- **Secure Restore**: This allows you to restore VMs that you don't want ransomware or malware to jeopardize. *Secure Restore* uses your antivirus program during the restore process to do complete scans of the VMs and files to ensure that there are no infections. If there are any infections or viruses, you can take action during the restoration process.

  Some of the default virus scanners supported are as follows:

  a) Symantec Protection Engine

  b) ESET

  c) Windows Defender

  d) Kaspersky Security 10

  Refer to the *Further reading* section for a link that talks in more detail about the antivirus scanners supported and how to configure them.

  The following diagram shows the steps that Veeam Backup & Replication takes during a *Secure Restore* process. You select a restore point, this is mounted, the virus scanner gets triggered to scan the mounted volumes, and if nothing is found, the restore operation continues. If a virus is found, it can be added to a secure network for remediation, or the process can terminate:

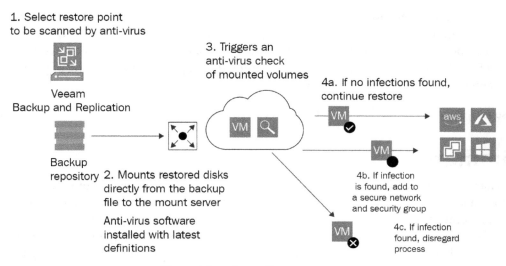

Figure 7.37 – Secure Restore process

- **Staged Restore**: This process allows you to inject a process during the restore of your VMs that can manage compliance-related requirements, such as those related to **GDPR** (short for **General Data Protection Regulation**).

The following diagram illustrates the *Staged Restore* process, where you can have a script to complete tasks that might be required on a VM to add security or change file permissions:

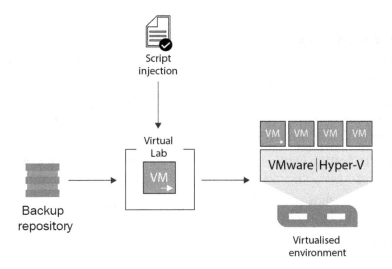

Figure 7.38 – Staged Restore process

- **Dev QA and Test**: Many times, developers and those managing servers wish to test application updates or patches prior to production rollout. Using the On-Demand Sandbox with the virtual lab will allow for this. You can spin up your VMs from a SureBackup job and leave the VMs running to complete manual testing of your applications and windows patching.

These are just some of the use cases that you can use with the Veeam DataLab features.

Summary

This chapter took a look into Veeam DataLabs and gave a brief explanation of what it is. We took a look at the components you need in order to set up complete validation testing and many other things using a Veeam DataLab. We reviewed how to configure the Veeam DataLab components – Virtual Lab, Application Group, and SureBackup jobs. A number of use cases that can work well in a Veeam DataLab were also discussed. After reading this chapter, you should now have a much deeper understanding of Veeam DataLabs and what they are. You should also have an understanding of the components that make up the Veeam DataLab and be able to configure them within Veeam Backup & Replication.

Hopefully, you will now have a better idea of the benefits of Veeam DataLabs. The next chapter, – *Cloud Backup and Recovery*, will take a deep dive into choosing the right **Managed Service Provider (MSP)**, ransomware, and cloud protection using RansomGuard, explain what RansomGuard is, and how you set it up from an MSP perspective.

Further reading

- On-Demand Sandbox for Storage Snapshots: https://helpcenter.veeam.com/docs/backup/vsphere/sandbox_storages.html?ver=100

- Authoritative versus Non-Authoritative Restore for Active Directory: https://www.veeam.com/blog/how-to-recover-a-domain-controller-best-practices-for-ad-protection.html

- Recovery Verification Options: https://helpcenter.veeam.com/docs/backup/vsphere/recovery_verification_overview.html?ver=100

- On-Demand Sandbox (DataLab): https://helpcenter.veeam.com/docs/backup/vsphere/sandbox.html?ver=100

- Antivirus Configurations for Secure Restore: https://www.veeam.com/blog/automate-recovery-verification-of-vm-backups-with-veeam-surebackup.html

- Secure Restore with DataLab: `https://helpcenter.veeam.com/docs/backup/vsphere/av_scan_about.html?ver=100`

- Custom Scripts for SureBackup: `https://www.veeam.com/blog/automate-recovery-verification-of-vm-backups-with-veeam-surebackup.html`

- Staged Restore with DataLab: `https://helpcenter.veeam.com/docs/backup/vsphere/staged_restore_about.html?ver=100`

8

Cloud Backup and Recovery Using Veeam Cloud Connect Provider and the Insider Protection Feature

Veeam Backup & Replication has several options regarding using a **service provider** for offsite backups for disaster recovery. When it comes to protecting your data, including the 3-2-1 rule, having offsite backups is very important to protect your data and ensure you have a way to recover if your onsite backups fail, and this chapter will cover this. This chapter will take a look at cloud backup and recovery and choosing a service provider. We'll discuss what ransomware is and how using a cloud provider can help your recovery.

You will learn about **Insider Protection**, which is a service provider feature enablement, learning what it is, and how it helps protect your data. Finally, we will dive into how you go about configuring Insider Protection within the Veeam Backup & Replication program to protect tenants' backups and data.

By the end of this chapter, you will have the required knowledge to research and, hopefully, choose a service provider to protect your data offsite. You will also have an understanding of Insider Protection, its benefits, and being able to request enablement from the service provider.

In this chapter, we're going to cover the following main topics:

- Exploring cloud backup and recovery – choosing a service provider
- Discovering what ransomware is
- Investigating and understanding cloud protection with a service provider – Insider Protection

Technical requirements

For this chapter, you should have Veeam Backup & Replication installed. If you've followed along with the book, then *Chapter 1*, *Installation – Best Practices and Optimizations*, covered the installation and optimization of Veeam Backup & Replication, which you can leverage in sections of this chapter.

Exploring cloud backup and recovery – choosing a service provider

When looking into how to leverage your backups with Veeam Backup & Replication, one of the additional benefits to complement your onsite backups would be choosing a **service provider** for other services. **Service providers** are third-party companies that can offer services such as **BaaS** (short for **Backup as a Service**) and **DRaaS** (short for **Disaster Recovery as a Service**). When looking into service providers, you should compare them based on a few things:

- **Services** – What specific services does the service provider offer? Is it only one of BaaS or DRaaS or potentially both? Depending on the provider, they could be a full-service shop where they will help you from start to finish and take care of you after the fact, while some only offer services for some of your backup and recovery needs. This is often overlooked by many users in complex restore scenarios. Service providers have many options at their disposal.

- **Security** – Will the service provider keep your data safe, and do they have best practices for security? You will want to ask about and understand their security procedures when it comes to handling your data.

- **Customer service** – This is one of the more essential aspects to look at for sales, account management, implementation, and aftercare support. Does the service provider offer all of these, and do they have client references you can speak to about their services?

- **Expertise** – Does the service provider know what they are doing and do they have the experts to take care of your systems and services appropriately? Having experienced professionals that have the required expertise is invaluable.

- **Customization** – Many service providers offer customized solutions to meet your requirements and needs. When looking, be sure to see if this is available as you might require something that is *out of the norm* of a typical offsite backup scenario, and a service provider that can meet your needs will be an asset.

- **Onboarding** – When you are looking at services, be sure to ask about things such as timelines and costs associated with getting you going with services. If you have a timeline or cost factor to meet, discuss this while selecting a service provider.

- **Cost** – One thing everyone looks at is the cost of backups on Microsoft Azure or Amazon S3. So, when choosing a service provider, look at the cost model they have for specific services to see if that meets your budget and project constraints.

So, as you can see, there are many factors to consider when you are looking at choosing a service provider for offsite backup or replication using the BaaS or DRaaS services. What precisely are these services that service providers offer? We'll take a look at their definitions and what each service involves when it comes to backups.

BaaS – Backup as a Service

BaaS is the approach used by service providers for a fully managed backup solution. The service provider manages your backups, restores, archiving, and so forth. However, there is an option where you can easily obtain a backup repository offsite, which gets done with the **Veeam Cloud Connect** product. Veeam Cloud Connect is used for cloud backups very easily:

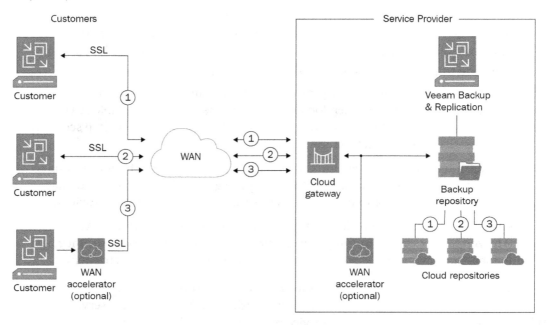

Figure 8.1 – Backup as a Service architecture

When looking at the BaaS service architecture, you can break it down further into a simplified diagram that shows the customer site and the service provider site and how the backup takes place using a backup copy job:

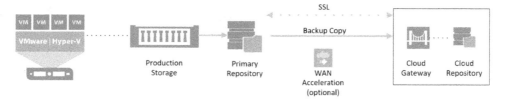

Figure 8.2 – BaaS overall concept

The main key component for both BaaS and DRaaS is adding a service provider to your console to access the service provider resources. To do this, you would complete the following:

1. Open the Veeam Backup & Replication console and click on the **BACKUP INFRASTRUCTURE** tab, where you will see the **Service Providers** section:

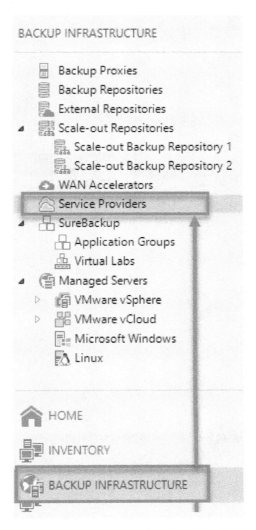

Figure 8.3 – Service Providers section of BACKUP INFRASTRUCTURE

2. After you click here, you can use the **Add Provider** button in the toolbar or right-click on the detail pane to the right to select **Add service provider**:

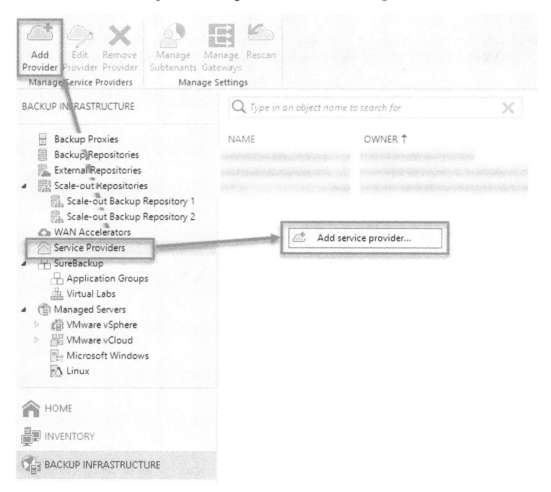

Figure 8.4 – Adding a new service provider

3.   Once you select one of the preceding options, you get presented with the **Service Provider** wizard:

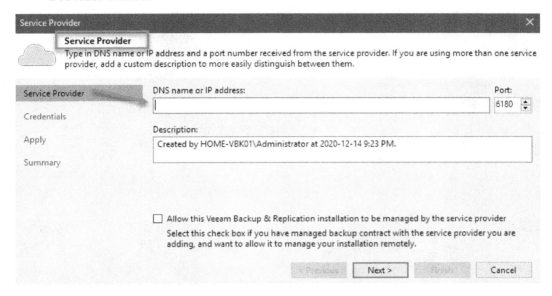

Figure 8.5 – Service Provider wizard

4. In this part of the dialog, you enter the **DNS name or IP address** of the service provider, the **Port** number (the default is 6180), and you have the option to select the checkbox to allow the service provider to be able to manage your backup server:

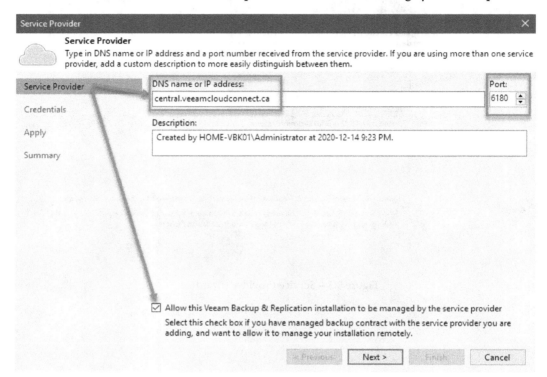

Figure 8.6 – DNS name, port, and checkbox selected

> **Important note**
>
> When selected, the checkbox allows the service provider to use another product called Veeam Service Provider Console to monitor and even manage your backups. If there are problems, support can connect to your backup server over a secure tunnel to help troubleshoot and fix jobs.

5. Once you have entered the details, you click **Next** to proceed to the next step, which is the **Credentials** for connection to the service provider, provided by them. Simultaneously, the SSL certificate used for the service provider is also validated, and if self-signed, you get prompted with another dialog to click **Continue**:

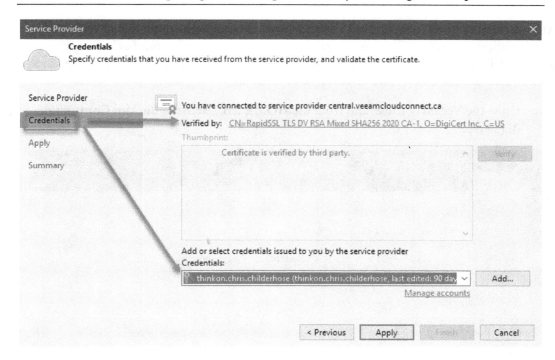

Figure 8.7 – SSL certificate verification and Credentials for Service Provider connection

6. Once you have the credentials selected, click on the **Apply** button to save the settings and then show the **Summary,** at which time you click **Finish** to complete the wizard:

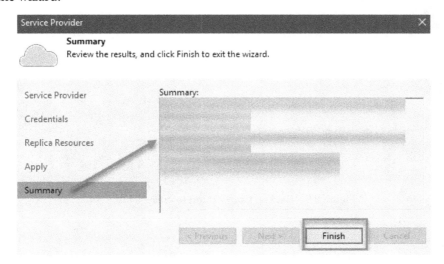

Figure 8.8 – Summary dialog

After clicking on the **Finish** button, a service provider scan gets run to update your configuration database with the new resources you can now use in your backup copy jobs. We will now walk through the process to set up a backup copy job that uses the BaaS service provider added to Veeam Backup & Replication:

1. Open the Veeam console, and from the **HOME** tab, select the **Backup Copy** option in the tree:

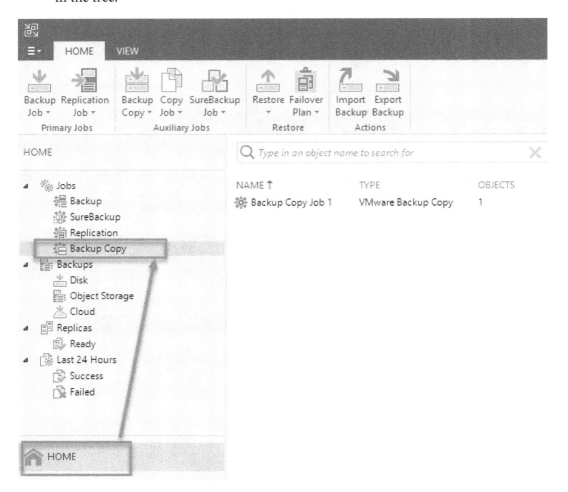

Figure 8.9 – Backup Copy option to create a job

2. Now click the **Backup Copy** button in the toolbar or right-click on the screen's right side and select **Backup Copy** to launch the new job wizard:

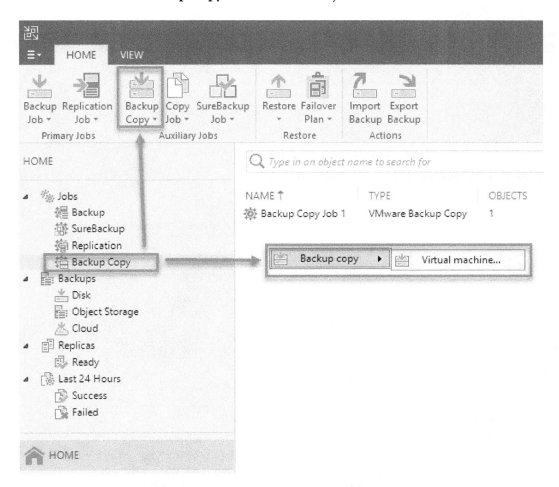

Figure 8.10 – Backup Copy option in the toolbar or right-click menu

3. In the **New Backup Copy Job** wizard, you need to enter a **Name, Description**, and then choose from two options – **Immediate copy (mirroring)** or **Periodic copy (pruning)**:

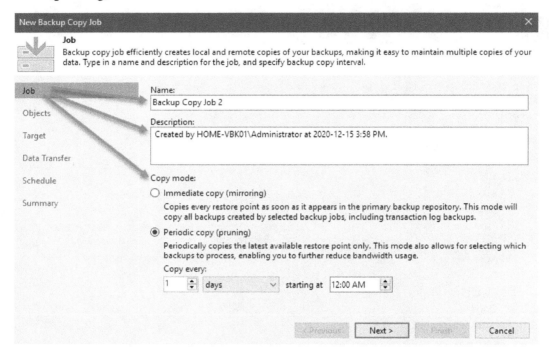

Figure 8.11 – New Backup Copy Job wizard

Important note

Immediate copy (mirroring) allows Veeam Backup & Replication to send your backups as soon as the primary backup job completes. In contrast, **Periodic copy (pruning)** copies it every so many days, at the specified time, as shown. Keeping the 3-2-1 rule in mind, the mirroring will create a restore point in the service provider target as soon as it can with the interval configuration. This is a great way to quickly get the same recovery points created on two different storage devices and in two different locations.

4. After your selections, click **Next** to proceed to the **Objects** part of the wizard. Here, you will select from infrastructure, backups, or jobs:

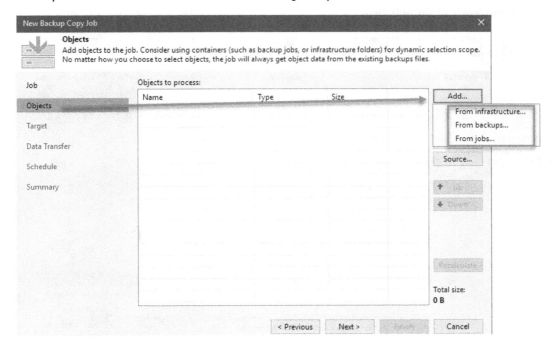

Figure 8.12 – Object selection for the copy job

The options in this dialog are as follows:

a) **From infrastructure** – Allows you to select servers from your virtual infrastructures such as vCenter or vCloud.

b) **From backups** – Allows you to select servers from backup jobs that you already have running.

c) **From jobs** – Lets you select the backup job to reference for the copy job.

5. Once you have selected the appropriate options, click on **Next** to proceed to the wizard's **Target** section, where you will specify your service provider settings you have added to the console:

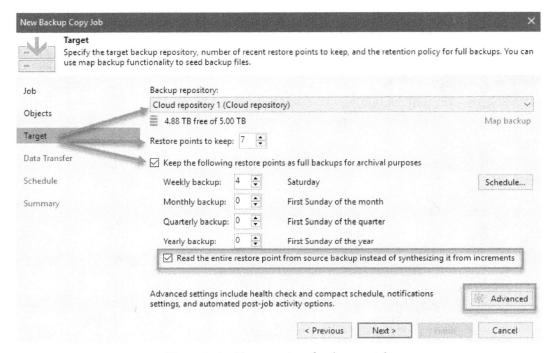

Figure 8.13 – Target options for the copy job

In this part of the wizard, you will see the cloud repository from the service provider you added to the console. You will also specify things such as the number of restore points to keep, and you can also set **GFS** (short for **Grandfather, Father, Son**) options as well using the checkbox **Keep the following restore points as full backups for archival purposes**.

> **Important note**
>
> The checkbox located just below the GFS section allows you to enable the backup copy job to read the entire restore point versus synthesizing it from the incremental backups, basically like doing a synthetic full backup. This is typically not enabled as a best practice because it would take longer to create the backup file for GFS.

6. Once you select all options, click **Next** to proceed to the **Data Transfer** section of the wizard, where you will specify **Direct** or **Through built-in WAN accelerators**:

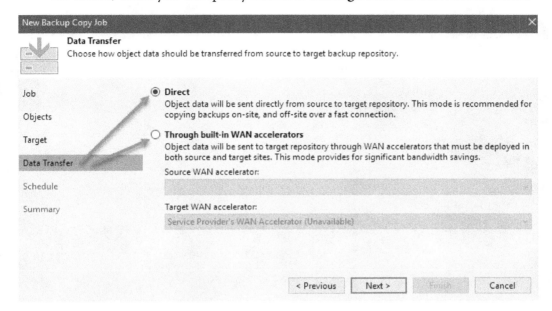

Figure 8.14 – Data transfer method selection

> **Tip**
> The **Through built-in WAN accelerators** option is useful when you have a slow internet connection as it will save bandwidth by compressing backups before sending them to the service provider. Be aware that WAN accelerators must be installed on both the user side and service provider side to work correctly.

7. Once you click **Next**, you are taken to the **Schedule** screen and can then click **Apply** and **Finish** to complete the setup:

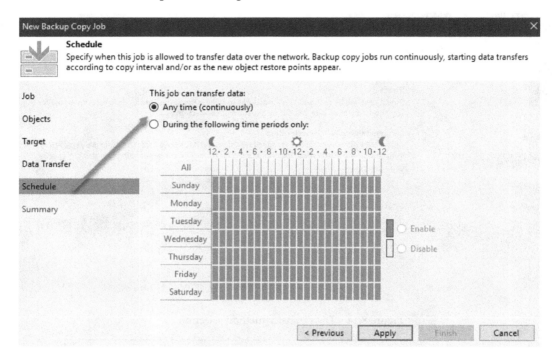

Figure 8.15 – Schedule settings for the copy job

Depending on your internet settings, you may not want to select the **Any time (continuously)** option and select a scheduled time range to run your backup copy job.

This section has covered what BaaS is and how to configure the console's service provider and a backup copy job. BaaS does have a limitation when it comes to recovering data should your onsite servers fail, and that is you need to rebuild everything, connect to the service provider, and then restore. The following section on DRaaS has the ability to recover your services in the service provider infrastructure to get you up and running while you work to rebuild your onsite services. Now let's quickly look at what DRaaS is and its use.

DRaaS – Disaster Recovery as a Service

DRaaS can replicate your servers to a service provider and then have them run on the infrastructure should a disaster happen. The following diagram illustrates the overall concept of DRaaS:

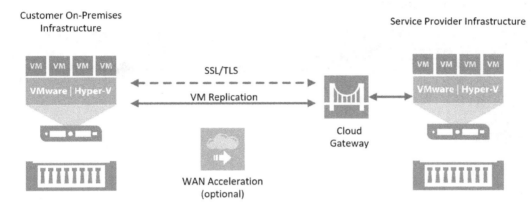

Figure 8.16 – Disaster Recovery as a Service (DRaaS)

DRaaS uses the replication jobs to replicate the servers from onsite to the service provider via Veeam Cloud Connect. Adding a service provider is the same as above, in the *BaaS – Backup as a Service* section. If you follow those steps to add the service provider, you then create a replication job directed to the service provider's cloud resources once that is complete.

The following steps outline how to set up a replication job to use DRaaS for the service provider:

1. Within the Veeam console and from the **Home** tab, click on the **Replication** option in the tree, and then use either the **Replication Job** button in the toolbar or right-click to select **Replication**:

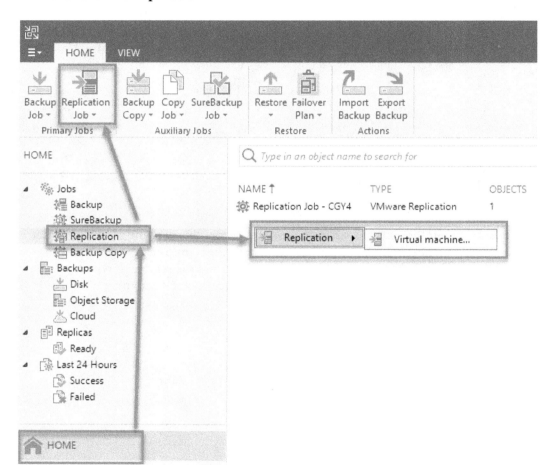

Figure 8.17 – Replication section for creating jobs

2. You are now in the **New Replication Job** wizard where you enter a **Name,
    Description**, and enable the advanced controls for **Replica seeding, Network
    remapping**, and **Replica re-IP**:

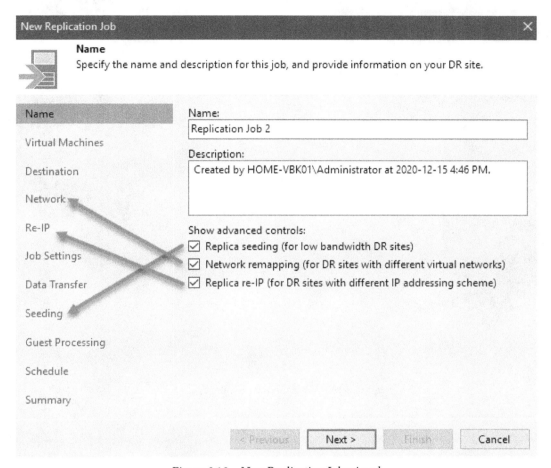

Figure 8.18 – New Replication Job wizard

3. Click **Next** to proceed to the **Virtual Machines** section of the wizard to select what you want to replicate using the **Add** button:

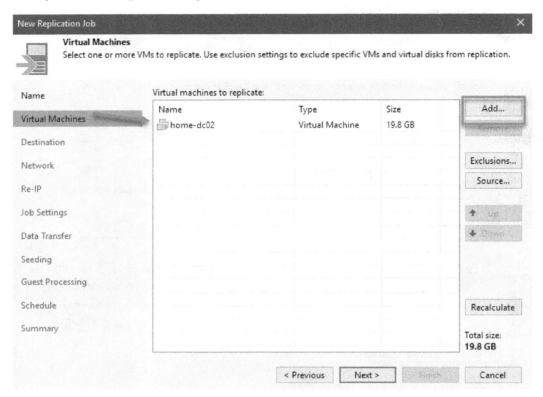

Figure 8.19 – Virtual machine selection

4. Once you select all your servers, click **Next** to continue to the **Destination** section of the wizard. You will choose the service provider settings that got added earlier by clicking the **Choose** button and selecting **Cloud host**:

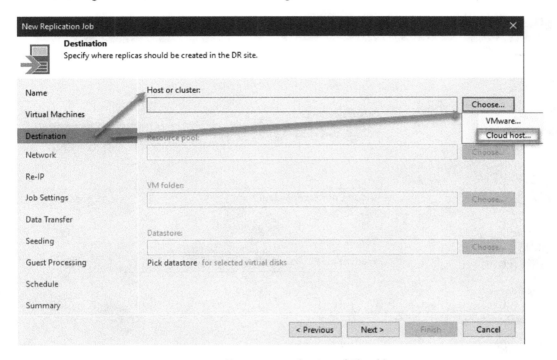

Figure 8.20 – Destination selection of Cloud host

5. Once you have selected the service provider, there is a potential for advanced options turning off, such as **Re-IP**. It is due to the settings configuration on the service provider side when creating the tenant in Veeam Cloud Connect:

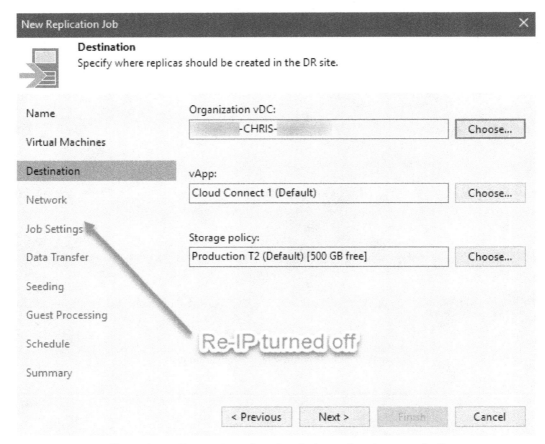

Figure 8.21 – Destination selected and advanced option turned off

6. Click **Next** to proceed to the wizard's **Network** settings where you configure your **Source network** and **Target network**:

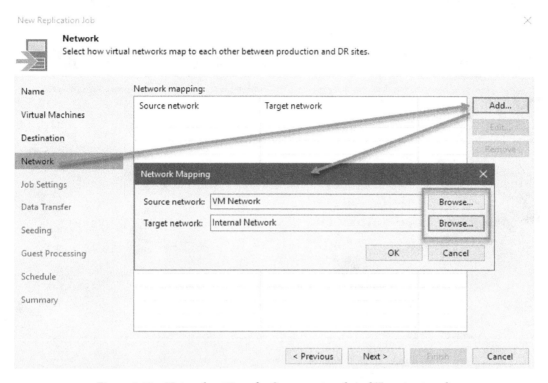

Figure 8.22 – Network settings for Source network and Target network

7. After selecting your networks, click **OK**. Then click **Next** to proceed to the **Job Settings** section of the wizard, where you will set things such as the repository for metadata (stored information about servers for replication) and the **Replica settings**:

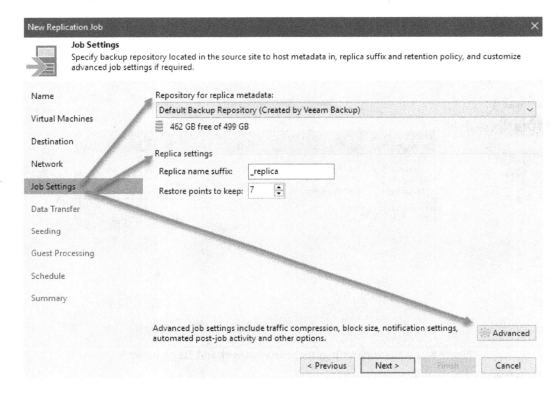

Figure 8.23 – Job Settings for replicas

8. Click **Next** to proceed to the **Data Transfer** section and set the proxy server and the transfer method – **Direct** or **Through built-in WAN accelerators**:

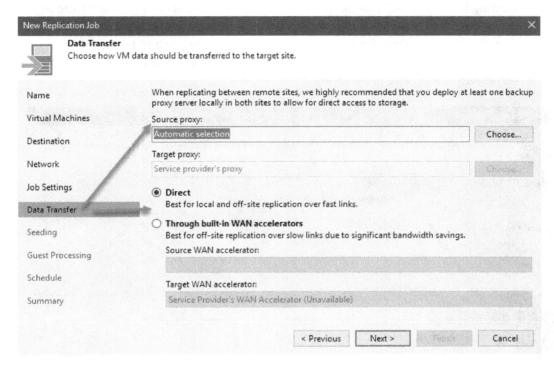

Figure 8.24 – Data transfer settings

> **Important note**
> To use WAN acceleration requires deployment on the user side and service provider side. Also, note that some options are grayed out and are set automatically by the settings of the service provider.

9.   Click **Next** to proceed to the **Seeding** tab if this got selected on the first screen of the wizard where you set the **Initial seeding** repository and **Replica mapping**. **Initial seeding** is the process of using backup data located at the service provider's site to create the initial replicas of your servers on their infrastructure. If you use seeding, then **Replica mapping** allows you to map the servers you wish to replicate to the ones used on the seeding repository, rather than creating a new replica copy of the server with the service provider:

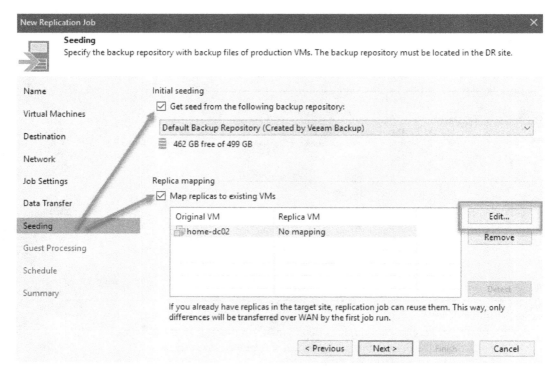

Figure 8.25 – Seeding replication settings

10. Click **Next** to proceed to the **Guest Processing** section, where you can turn on **application-aware processing**. This option allows crash-consistent backups:

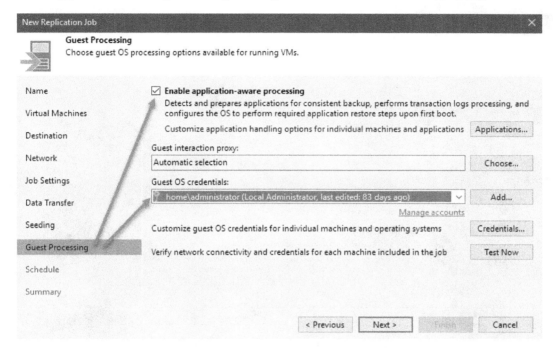

Figure 8.26 – Guest Processing settings

11. Click **Next** to proceed to the **Schedule** section, set the schedule options, and then click **Apply** and **Finish** on the **Summary** screen to create your replication job:

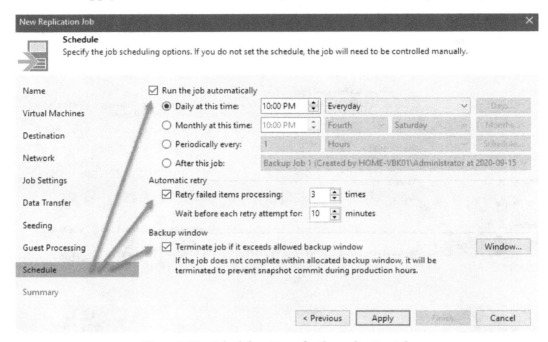

Figure 8.27 – Schedule settings for the replication job

We have now completed looking at DRaaS and how to create a replication job to use the service provider infrastructure for disaster recovery. Next, let's look at ransomware and how using both BaaS and DRaaS helps mitigate your data and servers' recovery.

Discovering what ransomware is

What is ransomware? **Ransomware** is defined as maliciously created malware that finds and encrypts an organization's files and storage. Many ransomware attacks are generated through phishing emails or other means to get within an organization's network to begin an attack.

Ransomware typically is used to extort money from organizations in the form of a ransom (hence the name ransomware) to receive the required keys to unencrypt files, or you would need to rely on backups to restore them. However, many ransomware attacks seek out backups and documents within a network to encrypt them, requiring the victim to pay the criminals.

Veeam Backup & Replication has many ways to protect your data against ransomware. We have discussed two of them in this chapter – BaaS and DRaaS. By utilizing these and some other capabilities built into Veeam Availability Suite, you can safeguard your data against attacks.

Other forms of ransomware capabilities in Veeam are utilizing Veeam Backup & Recovery along with **Veeam ONE**, another tool used to monitor your backups and infrastructure. Some of the features are as follows:

- **Ransomware Detection** – The Veeam ONE product has built-in monitors to detect anomalies that could be ransomware and alert you. It is worth noting that even Veeam Community Edition (the free version) has the Veeam ONE alarm for possible ransomware activity. At a minimum, everyone should be running this alert and detection mechanism for their VMware and Hyper-V environments.

- **Immutable Backup** – Used within the Capacity Tier of a SOBR to ensure backups cannot be deleted or modified.

- **Insider Protection** – Available with service providers, it allows them to turn on a recycle bin for your deleted backups that can be recovered based on the number of days configured.

- **DataLabs** – Allows you to configure isolated environments for conducting restores and not infecting your network.

- **Secure Restore** – This allows antivirus scanning of all files during the restore process so that you know you will not have ransomware or any other infections.

- **SureBackup** – Using SureBackup jobs, you can scan your backups with antivirus software to ensure that, if you need to restore, there are no infections.

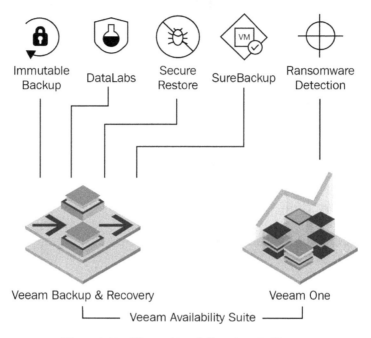

Figure 8.28 – Illustration of all options in Veeam

Noted in the preceding illustration, **Ransomware Detection** and **Veeam ONE** are discussed further in the book's final chapter – *Chapter 10, Veeam ONE*. Now that we understand ransomware and what it is, let's look at how a service provider can help you mitigate this even further using a technology built into *Veeam Cloud Connect* called **Insider Protection**.

Investigating and understanding cloud protection with a service provider – Insider Protection

When you are investigating service providers, you will want to ask whether they offer *Insider Protection* or not? This is important because it adds another layer of security on top of the BaaS or DRaaS offerings to protect your data further from ransomware.

In some cases, having primary or additional backups in a service provider cloud repository will not be enough to ensure data security for your tenants. The data could become unavailable due to an insider attack. Say a hacker has gained access to the Veeam Backup & Replication console and deleted all the tenant's backups, including those that are offsite. Another case could be an administrator who accidentally or intentionally deletes backups from the cloud repository. Insider Protection will help protect users against these types of attacks.

For a service provider to allow you to access Insider Protection, you need to enable a checkbox located within the tenant settings in the **Backup Resources** section. You select the checkbox and set the number of days to keep the backups in the *recycle bin*, which is discussed next. Once this is complete, if a user deletes backups accidentally or intentionally, or has a ransomware attack, these can be restored by the service provider:

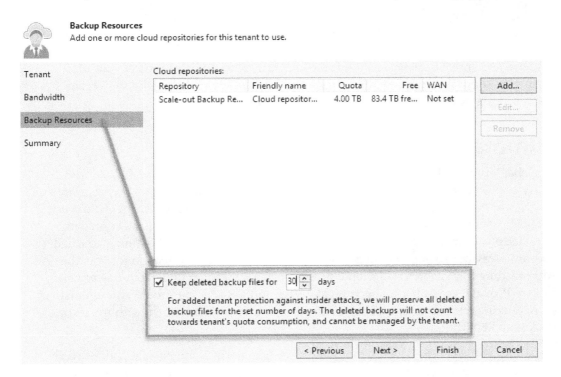

Figure 8.29 – Insider Protection option enabled

With this option enabled, when a backup or a specific restore point in the backup chain gets deleted from the cloud repository, Veeam Backup & Replication does not immediately delete the actual backup files. Instead, Veeam Backup & Replication will move backup files to the *recycle bin*.

The *recycle bin* is a folder created on the backup repository in the service provider infrastructure. The folder gets automatically created when the tenant backups get moved to the *recycle bin* and the first backup files located in the *recycle bin* do not consume any of the tenant's storage quota. However, they will consume disk space in the service provider storage infrastructure where the cloud repository resides. When a service provider offers Insider Protection, they need to account for this and provide enough storage capacity.

Backup files moved to the *recycle bin* will appear deleted to the tenant. To gain access to the backup files, they must contact the service provider to restore them from the *recycle bin* back to the tenant's repository folder. At that time, the tenant will need to rescan the repository to access the restored backups.

> **Important note**
> The backups kept in the *recycle bin* for <*N*> days get permanently removed once they expire and will not be accessible after that.

We have now looked at what Insider Protection is and how it helps protect your data against ransomware attacks. We also looked at how to configure it within the Veeam Backup & Replication console for the tenant settings.

Summary

This chapter took a look into cloud backup and recovery using BaaS and DRaaS with a service provider. We looked at how to investigate service providers and their offerings as well as what to look for. We reviewed what ransomware is and how you can avoid it using BaaS, DraaS, and Insider Protection. We also discussed using Insider Protection with a service provider to add an extra layer of security to your backups. We finally touched on how a service provider would go about configuring Insider Protection for you, the user. After reading this chapter, you should now have a much deeper understanding of what Service Providers can offer for services like BaaS and DRaaS. You should also understand these offerings as well as Insider Protection and how to configure them within Veeam Backup & Replication.

Hopefully, you will now have a better idea of the benefits of cloud backup and recovery when it comes to protecting your data. The next chapter – *Instant VM Recovery* – will dive into how this process works within Veeam Backup & Replication, and when you should use it.

Further reading

Here are links to more information on some of the topics that we have covered in this chapter:

- Connecting to service providers from Veeam console: `https://helpcenter.veeam.com/docs/backup/cloud/cloud_connect_sp.html?ver=100`

- BaaS and DRaaS information for Veeam: `https://www.veeam.com/blog/cloud-availability-options-baas-draas.html`

- Backup copy job information: `https://helpcenter.veeam.com/docs/backup/vsphere/backup_copy.html?ver=100`

- Backup copy job best practices: `https://bp.veeam.com/vbr/VBP/4_Operations/O_Veeam_Jobs/O_backup_copy_jobs/backup_copy_job.html`

- Ransomware and Veeam: `https://www.veeam.com/ransomware-protection.html`

- Veeam ransomware papers: `http://vee.am/ransomwareseriespapers`

- Veeam Cloud Connect Insider Protection: `https://helpcenter.veeam.com/docs/backup/cloud/cloud_connect_bin.html?ver=100`

- Replication job information: `https://helpcenter.veeam.com/docs/backup/qsg_vsphere/replication_job.html?ver=100`

- Replication job best practices: `https://bp.veeam.com/vbr/VBP/4_Operations/O_Veeam_Jobs/O_replication_jobs/replication_jobs.html`

9
Instant VM Recovery

Veeam Backup & Replication has several options regarding restoring a **virtual machine (VM)**, and one of those is using the **Instant VM Recovery** feature. In this chapter, you will take a look at Instant VM Recovery and how to use it. We'll then discuss what Instant VM Recovery is when it comes to restoration. You will also learn the prerequisites and requirements for conducting an Instant VM Recovery, before starting and completing an Instant VM Recovery. Finally, we will dive into how you go about committing your Instant VM Recovery by migrating it to your environment or canceling the recovery process.

By the end of this chapter, you will be able to explain what Instant VM Recovery is, discovered what you need in order to complete the recovery process, how to define the steps from start to finish in an Instant VM Recovery, and how to migrate your servers to your production environment or cancel the recovery process.

In this chapter, we're going to cover the following main topics:

- Exploring Instant VM Recovery – what is it?
- Discovering the requirements and prerequisites for Instant VM Recovery
- Investigating and understanding the Instant VM Recovery process and steps
- Exploring the migration or cancelation steps for recovery

Let's get started!

Technical requirements

For this chapter, you should have Veeam Backup & Replication installed. *Chapter 1, Installation – Best Practices and Optimizations*, covers how to install and optimize Veeam Backup & Replication, which you can leverage in this chapter.

Exploring Instant VM Recovery – what is it?

Within Veeam Backup & Replication, there is your typical backup and restore operations, but what happens when you need to recover an entire VM that has corrupted files or files that have been accidentally deleted? This problem is where the *Instant VM Recovery* option comes in handy – it allows you to immediately restore different workloads as **VMware vSphere** VMs by running them directly from your backup files. The supported backup file types are as follows:

- Backups of **VMware vSphere** VM created with Veeam Backup & Replication

- Backups of **vCloud Director** virtual machines created with Veeam Backup & Replication

- Backups of **Microsoft Hyper-V** virtual machines created with Veeam Backup & Replication

- Backups of virtual and physical machines created with **Veeam Agent for Microsoft Windows** or **Veeam Agent for Linux**

- Backups of **Nutanix AHV** virtual machines created with Veeam Backup for Nutanix AHV

- Backups of **Amazon EC2** instances created with Veeam Backup for AWS

- Backups of **Microsoft Azure** virtual machines created with Veeam Backup for Microsoft Azure

As you can see, there are many options for restoring workloads using Instant VM Recovery and backups from various sources. With Instant VM Recovery, you can help improve **recovery time objectives** (**RTO**) and minimize the disruption and downtime of production workloads. Using Instant VM Recovery is like you have a *temporary spare* for your workload. Users remain productive instead of allocating unestimated amounts of time troubleshooting any issues with the failed workload.

When you perform an Instant VM Recovery, Veeam Backup & Replication uses the **Veeam vPower** technology to directly mount the image on an ESXi host from the compressed and deduplicated backup file for the VM. Since no extraction is necessary from the backup file, this allows you to select any restore point for a quick restore in a matter of minutes.

When you select a restore point for Instant VM Recovery, the image remains in a read-only state to avoid unexpected modifications. Instead, all changes to the virtual disks are logged in an auxiliary redo log file that resides on the NFS server (backup server or backup repository) using the **vPower NFS** service. These recorded changes either get discarded when the restored VM is removed or are merged with the original VM data if recovery gets finalized and moved to production.

I/O performance for a restored VM is improved when it's redirected to a specific datastore during the Instant VM Recovery process. In this case, however, instead of using redo logs, Veeam Backup & Replication triggers a snapshot. It puts it into the **Veeam IR** directory on the selected datastore, together with the metadata files holding the VM's changes.

There are three options when it comes to finalizing your Instant VM Recovery:

- **Storage vMotion**: This allows you to quickly migrate the restored VM to production storage with no downtime. Data gets pulled from the NFS datastore to the production storage, and changes are consolidated with the VM still running.

> **Important note**
> Storage vMotion can only be used if you choose to keep the VM changes on the NFS datastore without redirecting them, as noted previously for I/O performance. Storage vMotion is also only available with select licences of VMware.

- **Replication or VM Copy**: This allows you to create a copy of the VM and fail it over during a maintenance window or the next available opportunity. Unlike Storage vMotion, this option requires downtime to clone or replicate the VM, and then power it off and then power on the cloned copy or replica.

- **Quick migration**: This is a two-stage migration process that Veeam Backup & Replication will perform. Rather than pulling data from the vPower NFS datastore, it will restore the VM from the backup file on the production server and then move all changes and consolidate them with the VM data. Please see the *Further reading* section at the end of this chapter for a link to more information on this topic.

The overall process looks similar to this:

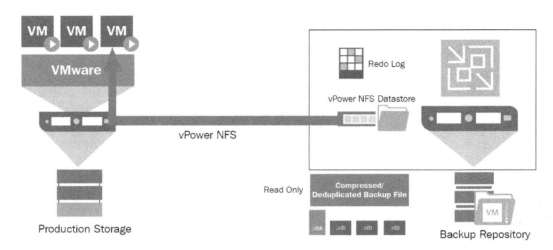

Figure 9.1 – Instant VM Recovery – overall steps

As noted in the preceding diagram, once you've selected the VMs you wish to recover, Veeam will read the backup files from the repository, send them to the server acting as the vPower NFS, and mount this to the ESXi host that was selected in the wizard. The redo logs that store the changes are then created either on the same NFS datastore or an alternate one specified in the wizard.

There is also another option other than Instant VM Recovery, and that is **Instant VM Disk Recovery**. What is the difference, you may be asking? Rather than recovering the entire VM, you can use Instant VM Disk Recovery to recover a single disk from a VM. Why would you want to do this? Well, there are two good reasons:

- Recover a VM disk, not the entire VM. If you need the whole VM, then you should use **Instant VM Recovery**.

- Recover a VM disk and keep the target VM (that you want to attach the disk to) powered on. However, if the VM can be powered off, you should use the **Virtual Disks Restore** option. The *Further reading* section contains a link to this option.

Now that we have taken a look at what Instant VM Recovery is and how it works, we will investigate the requirements and prerequisites for completing this process.

Discovering the requirements and prerequisites for Instant VM Recovery

There are a few components from Veeam Backup & Replication that are necessary to take into account. Storage space also needs to be considered:

- **vPower NFS service**: This service should be installed on a Windows server, which allows it to act as an NFS Server for mounting the backup files for the Instant VM Recovery process.

- **Disk space for cache folder**: Instant VM Recovery requires additional free space for the cache folder (non-persistent data, at least 10 GB recommended).

- **Production storage**: If you are going to use Storage vMotion or Quick Migration, then you will need enough storage for the VM to restore based on the size of the VMDK files.

Instant VM Recovery's main component is the *Veeam vPower NFS service* since it performs the bulk of the workload during the entire process. As noted previously, this is the Windows service that gets installed on any server to allow it to act as an NFS server that mounts the VM files from the backup files.

The Veeam vPower technology enables the following features:

- Recovery verification
- *Instant VM Recovery*
- *Instant VM Disk Recovery*
- Staged restore
- **Universal Application-Item Recovery (U-AIR)**
- Multi-OS file-level restore

The system designated as the vPower NFS server creates a particular directory called the **vPower NFS datastore**. When you start a VM or VM disk from backup, Veeam Backup & Replication "publishes" the VMDK files of the VM from backup on the vPower NFS datastore. The VMDK files are emulated (pointers get created) on the vPower NFS datastore, but the actual VMDK files remain on the backup repository.

Once the vPower NFS datastore creates the pointers to the VMDK files on the backup repository, it gets mounted to the ESXi host. The ESXi can now *see* the backed up VM images within the vPower NFS datastore and work with them as regular VMDK files. The emulated VMDK files function as pointers to the real VMDK files in the backup repository.

> **Important note**
>
> Veeam vPower NFS datastores are service datastores that can be used for vPower operations only. You cannot use them as regular VMware vSphere datastores. As an example, you can't place files of replicated VMs on such datastores.

When determining the vPower NFS Server service's location, it is strongly recommended that you enable it on the server acting as your repository on Windows. If your repository server is on Linux, then you will need a Windows server to install the service. Installing it directly on the repository server allows Veeam Backup & Replication to directly connect the backup repository and ESXi that the vPower NFS datastore gets mounted to. When the vPower NFS Server service is not running on the repository server, it can affect recovery verification performance. This is due to the connection between the ESXi host and backup repository getting split into two parts:

- From the ESXi host to the vPower NFS server
- From the vPower NFS server to the backup repository

The final thing to note about the vPower service is that you must ensure that a proper network interface configuration has been set up on the ESXi host and vPower NFS server. The ESXi host must access the vPower NFS server; otherwise, it won't be able to mount the datastore properly, if at all. The ESXi host uses the VMkernel interface to mount the vPower NFS datastore, so it must be present; otherwise, the vPower NFS datastore will fail to mount.

> **Important Note**
>
> Veeam Backup & Replication uses IP address authorization to restrict access to the vPower NFS server. By default, the vPower NFS server can only be accessed by the ESXi host that provisioned the vPower NFS datastore. However, you can disable this with a registry key that requires you to contact Customer Support for assistance.

You should now understand the Instant VM Recovery requirements, including the vPower NFS server service, that mount and communicate with the repository server and backup files. We will now dive into how the process works and how to run an Instant VM Recovery.

Investigating and understanding the Instant VM Recovery process and steps

When it comes to conducting an Instant VM Recovery, four main steps take place in the background:

1. **Initialization Phase**: This is the initial step in the process, and Veeam Backup & Replication prepares the resources necessary for Instant VM Recovery. It does the following:

   a) Starts the Veeam Backup Manager process on the Veeam backup server

   b) Checks the Veeam Backup service for whether the necessary backup infrastructure resources are available for Instant VM Recovery

   c) Communicates with the Transport Service on the backup repository to start the Veeam Data Mover

2. **NFS Mapping**: Once the backup infrastructure resources are ready, an empty NFS datastore is mapped to the selected ESXi host using the vPower NFS service. Veeam Backup & Replication then creates VM configuration files and links to the virtual disk files while remaining in the backup repository. Then, changes get written to the cache file.

3. **Registering and Starting VM**: The VM will run from the Veeam NFS datastore, which gets treated as any regular datastore in VMware vSphere that's attached to the ESXi host. For this reason, you can perform all actions vCenter Server/ESXi supports for regular VMs.

4. **Migrating guest to production datastore**: Once you have confirmed the VM is working as expected, you can then migrate the VM disk data to a production datastore. You can do this using either Storage vMotion or Veeam Quick Migration.

There are also a couple of performance concerns to note regarding the Instant VM Recovery process concerning its read patterns and write redirections:

- **Read pattern**: Instant VM Recovery is very read-intensive, and performance is directly related to the performance of the underlying repository. Good results can be obtained from standard drive repositories, while deduplication appliances should be considered carefully for such use.

- **Write redirections**: Once the guest VM has been booted, it will read existing blocks from the backup storage and write/reread new blocks on the configured storage, regardless of whether a datastore or a temporary file is on the vPower NFS server local drive, in the `%ALLUSERSPROFILE%\Veeam\Backup\NfsDatastore` folder. To ensure there's consistent performance during this process, Veeam recommends redirecting the writes of the restored guest to a production datastore.

To start an Instant VM Recovery, you need to consider the following:

- You can restore a workload from a backup that has at least one successfully created restore point.

- If you are restoring a workload to the production network, make sure that the original workload is powered off.

- If you want to scan restored VM data for viruses, check the secure restore requirements and limitations (see the *Further reading* section).

- You must provide enough free disk space in the vPower NFS datastore. The minimum amount of free space must be equal the RAM capacity of the recovered VM, plus 200 MB. For example, if the restored VM has 32 GB of virtual RAM, 32.2 GB of free space is required.

- By default, the vPower NFS datastore is located in the IRCache folder on a volume with a maximum amount of free space. When you choose to redirect virtual disk updates to a VMware vSphere datastore during job configuration, this location does not get used.

- **For the Veeam Quick Migration with Smart Switch**: You need to provide more disk space in the vPower NFS datastore, including the minimum, which is equal to the RAM capacity of the VM.

- **For Nutanix AHV VMs**: Instantly restored VMs will have default virtual hardware of two CPU cores, 4 GB of RAM, and one network adapter.

Now, let's go through the process of starting an Instant VM Recovery. Complete the following steps:

1. Launch the Instant VM Recovery Wizard by highlighting **Backups** on the **Home** tab and clicking the **Restore** button so that you can select **VMware vSphere**:

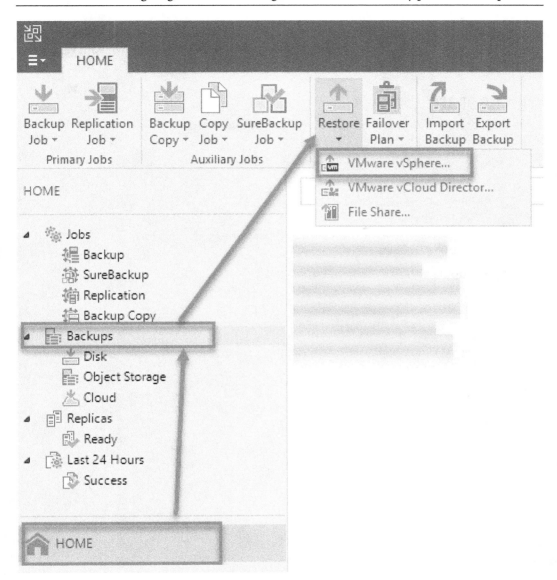

Figure 9.2 – Restore button to launch the wizard

2. Upon you clicking the **VMware vSphere** option, you will be presented with a **Restore** dialog asking if you want to **Restore from backup** or **Restore from replica**:

Figure 9.3 – Restore option for backup or replica

3. Click the **Restore from backup** option. You will be presented with the **Restore from Backup** wizard. From here, select the **Entire VM restore** option:

Restore from Backup
Select the type of restore you want to perform.

 Entire VM restore
Restores the entire VM or its individual components, such as virtual disks.

 Volume restore
Restores the content of individual volumes.

 Guest files restore
Restores individual guest files from an image-level backup.

 Application items restore
Restores individual application items from an image-level backup.

Cancel

Figure 9.4 – Entire VM restore selection

4. After clicking **Entire VM restore,** you will be presented with the option to select **Instant VM Recovery**, which launches the **Instant VM Recovery**:

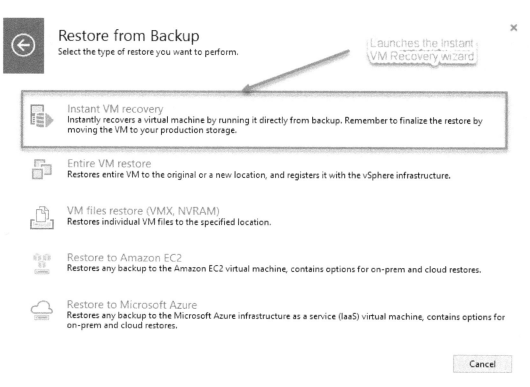

Figure 9.5 – Instant VM Recovery option to launch the wizard

5. Once you click the **Instant VM recovery** option, this will launch the **Instant VM Recovery** wizard, which is titled **Instant Recovery to VMware**:

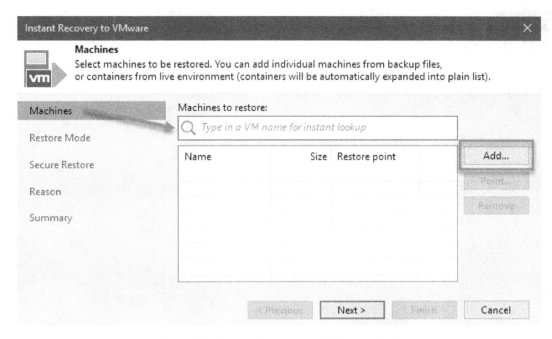

Figure 9.6 – Instant Recovery to VMware wizard

6. In the dialog box, type in a VM name to search for it or click the **Add** button to select a VM from **Backup**:

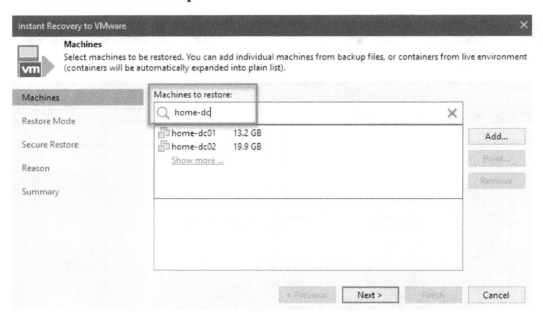

Figure 9.7 – Quick search by typing in the VM's name

You can also click **Add** to select your VM from **Infrastructure** or **Backup**:

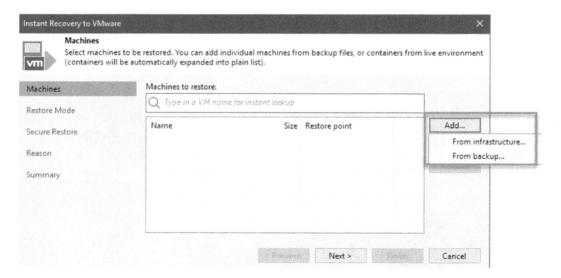

Figure 9.8 – Add button for selecting from Infrastructure or Backup

Also, note that one of the enhancements in v10 for Veeam Backup & Replication is conducting Multi-VM instant recovery. You can select more than one VM if you wish the wizard to perform this type of recovery.

7. Once you've selected your VM, click **Next** to proceed to the **Restore Mode** section of the wizard:

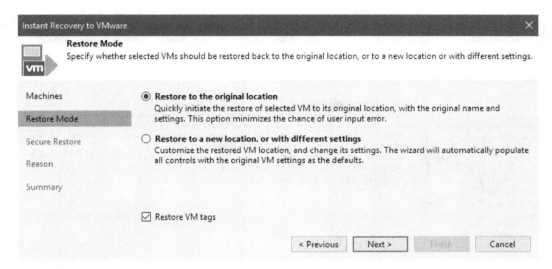

Figure 9.9 – Restore mode selection

8.   Now, select **Restore Mode** for **Restore to the original location** or **Restore to a new location, or with different settings**. Selecting the first option restores without any user input and takes you directly to the **Reason** section, whereas the second option allows you to choose from three options: **Destination**, **Datastore**, and **Secure Restore**. You can also select to restore the VM tags with the **Restore VM tags** checkbox:

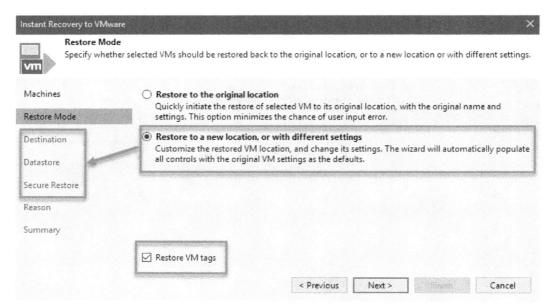

Figure 9.10 – Restore to a new location, or with different settings

9. Click **Next** to proceed to the **Destination** section of the wizard. Here, you can specify things such as **Restore VM Name**, **Host**, **VM folder**, and **Resource pool** and customize the UUID of the virtual machine with the **Advanced** button:

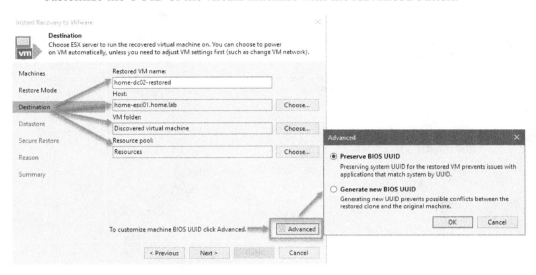

Figure 9.11 – Destination options and UUID

10. Once you have set all the required options, click **Next** to proceed to the **Datastore** section:

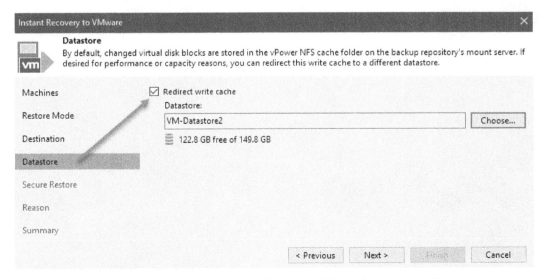

Figure 9.12 – Datastore section of the wizard for Redirect write cache

11. At this stage, you can select **Redirect write cache** to use a datastore instead of the vPower NFS cache folder on the server, which is acting as the vPower NFS server. As we mentioned previously, this can be for capacity or performance reasons. Once selected, click **Next** to proceed to the **Secure Restore** section:

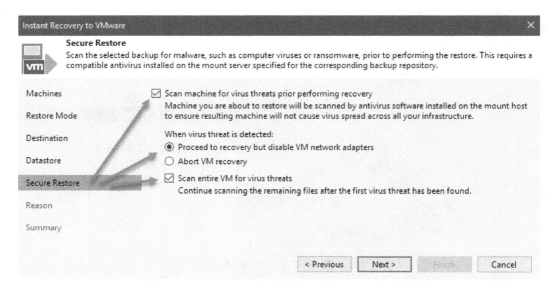

Figure 9.13 – Secure Restore options for scanning the VM for viruses during the restore process

12. At this stage, you turn on **Scan machine for virus threats prior performing recovery** and then choose to proceed or abort if anything is found. You can also select **Scan entire VM for virus threats**, which the VM(s) will take longer to restore and ensure there are no server infections. Once the options you prefer are enabled, click **Next** to proceed to the wizard's **Reason** section:

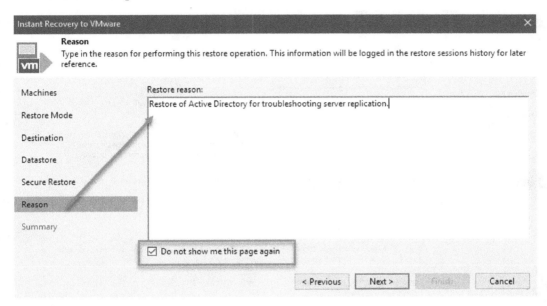

Figure 9.14 – Reason section for description and optional checkbox

13. Once you've entered a **Restore reason**, you will have the option to not show this page again by checking the relevant checkbox, if desired. Click **Next** to move on to the **Summary** page, and then **Finish** to begin the Instant VM Recovery job:

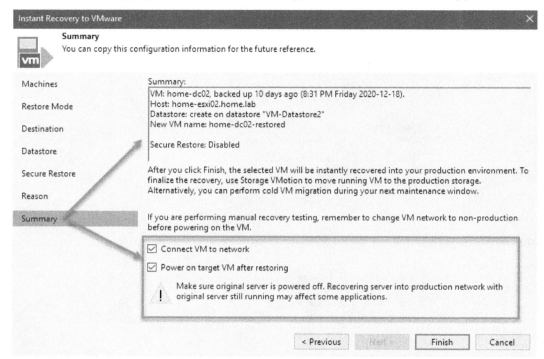

Figure 9.15 – Summary section and options for network and power

> **Important note**
> You can select two options here: **Connect VM to network** and **Power on target VM after restoring**. Note that you may or may not want to connect the network at first to avoid conflict with the original VM, but if you do select to power on the VM, a warning will appear to ensure the original VM is powered off, as noted in the preceding screenshot.

In the *Discovering the requirements and prerequisites for Instant VM Recovery* section, we discussed the vPower NFS server and how it mounts a datastore to the host that you chose to restore to in the wizard. The following is the datastore mounted to the host – home-esxi02.home.lab – for the Instant VM Recovery process:

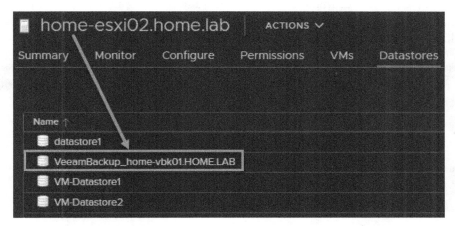

Figure 9.16 – vPower NFS mounted datastore for Instant VM Recovery

Remember that, in the wizard, the datastore named VM-Datastore2 was selected for the write cache change metadata.

You can now monitor the restoration process and watch your virtual environment to see the VM get recreated in the infrastructure. Next, we will learn how to place the VM into production or cancel/delete the restored VM.

Exploring the migration and cancelation steps for recovery

Now that we have gone through the **Instant VM Recovery** wizard so that we can begin the recovery process, we can see that the job is running and showing a status of **Mounted**. At this point, we can choose what to do with the VM in regards to keeping it in production or canceling the restore so that we can remove it:

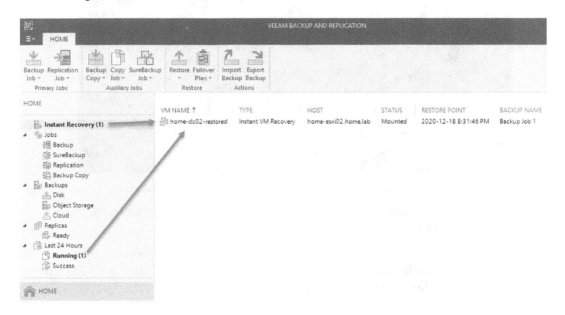

Figure 9.17 – Instant VM Recovery job in running state

When you select the job in the right-hand pane, the toolbar changes to give you four options:

- **Migrate to production**: This will launch a wizard that will start **Quick Migration** of the VM to move it to production.

- **Open VM Console**: This launches the VM console so that you can log into the VM and check things before migrating to production. This option will prompt you for your vCenter credentials.

- **Stop Publishing**: As its name suggests, it will unmount the datastore, remove the VM, and stop the Instant VM Recovery job to cancel everything.

- **Properties**: This shows the properties of the Instant VM Recovery job.

The following steps outline the **Migrate to production** option, when selected:

1. Click on the **Migrate to production** button or right-click on the job to select the same option:

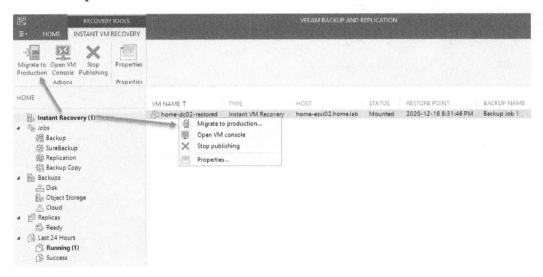

Figure 9.18 – Migrate to Production button or right-click the menu

2. You will then see the **Quick Migration** wizard, where you can input the
   Destination options for **Host or cluster**, **Resource pool**, **VM folder**, and **Datastore**:

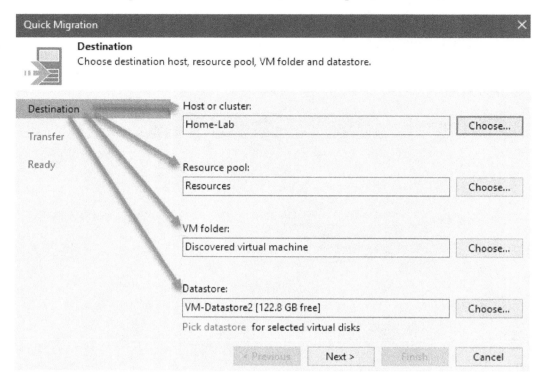

Figure 9.19 – Destination selection for Quick Migration

3. Once you have made your selections, click **Next** to proceed to the **Transfer** section
   of the wizard:

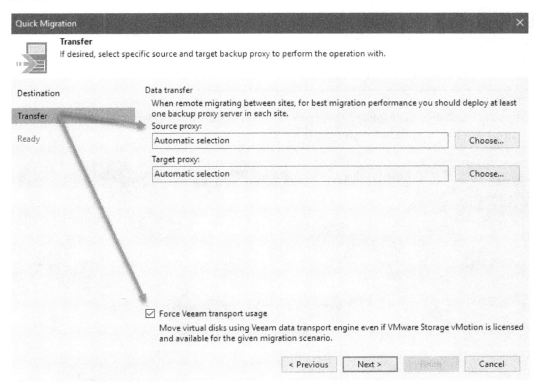

Figure 9.20 – Transfer options for Source/Target proxy and Veeam transport

4. Here, you can select specific **Source/Target proxy** servers to use and check the checkbox for **Force Veeam transport usage**. This option will work well if you have a license for VMware vSphere that does not have Storage vMotion to ensure the VM is transferred by Veeam Backup & Replication to your infrastructure. Otherwise, it is best to let VMware handle the vMotion of the VM (leave this unchecked). Click **Next** to proceed to the **Ready** section and click **Finish** to begin migrating your VM to production.

The **Quick Migration** wizard will run, and VMware vSphere will Storage vMotion your VM to the datastore selected before registering it on a host – if the cluster was chosen – or on the actual host specified in the wizard.

Clicking the **Open VM Console** button launches the console to the VM within the vCenter environment. You will be prompted for your vCenter credentials to log in, and then the console window will open so that you can log into the VM to validate it before migration to production or canceling it.

If you do not wish to migrate the VM to production, then you can click the **Stop Publishing** button in the toolbar. This option will remove the VM from your inventory and dismount the NFS datastore on the host. The job will be canceled:

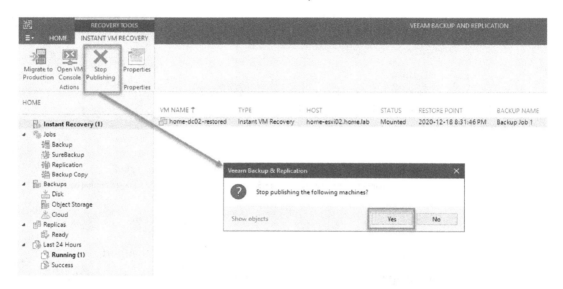

Figure 9.21 – Stop Publishing option to cancel the restore

Upon clicking **Yes**, the VM is dismounted from the datastore and completes:

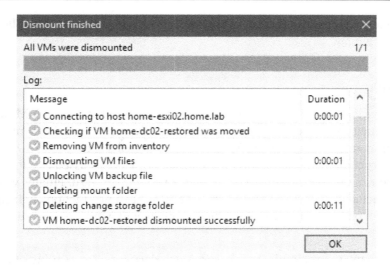

Figure 9.22 – Dismounting the VM and datastore – Stop Publishing

Clicking the **Properties** button brings up the restore job's details, which shows it is **In progress**:

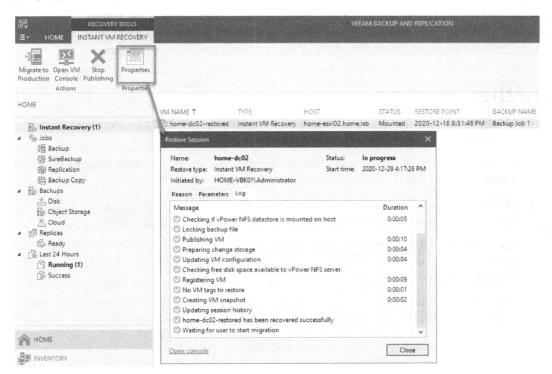

Figure 9.23 – Properties for Instant VM Restore job

With that, we have wrapped up the entire Instant VM Recovery process. Now, let's summarize this chapter.

Summary

In this chapter, we looked at Instant VM Recovery, including what Instant VM Recovery is and what backups you can use for this process. We reviewed the requirements and prerequisites for conducting an Instant VM Recovery, as well as how the vPower NFS Service plays a crucial role in performing these tasks. We also discussed and looked at the entire process of doing an Instant VM Recovery using the wizard. Finally, we touched on how to conduct a Quick Migration to move the VM into a production environment, as well as how to stop publishing to cancel the restore overall, which dismounts the VM and datastore.

By reading this chapter, you should now have a much deeper understanding of Instant VM Recovery and what it is. You should also understand the requirements and prerequisites for doing a recovery, including the vPower NFS service. Finally, you should understand how to migrate your VM to production or cancel and dismount both the VM and datastore. Hopefully, you now have a better idea of how Instant VM Recovery works and how easy it is to move your VM into production or simply cancel the restore. The next and final chapter, *Veeam ONE*, will dive into Veeam's monitoring and reporting tool, which can help you keep track of both your backup and virtual infrastructure.

Further reading

- Quick migration: `https://helpcenter.veeam.com/docs/backup/vsphere/quick_migration.html?ver=100`

- Virtual Disk Restore: `https://helpcenter.veeam.com/docs/backup/vsphere/virtual_drive_recovery.html?ver=100`

- (VIDEO) How to perform Instant VM Recovery: `https://www.veeam.com/videos/perform-instant-vm-recovery-15236.html`

- Secure Restore Requirements and Limitations: `https://helpcenter.veeam.com/docs/backup/vsphere/av_scan_about.html?ver=100#av_limitations`

- vPower NFS Information: `https://helpcenter.veeam.com/docs/backup/vsphere/vpower_nfs_service.html?ver=100`

10
Veeam ONE

Veeam Backup & Replication has a companion application that can monitor your virtual infrastructure and backup servers called **Veeam ONE**. In this chapter, we will take a look at Veeam ONE and how to use it. We'll discuss its installation requirements and the steps we need to follow to complete set up the program. You will then learn about how to monitor different environments such as vSphere, vCloud, and Veeam, before looking at the reporting features of Veeam ONE. Finally, we will dive into how to use Veeam ONE to troubleshooting things in an environment.

Veeam ONE is a great monitoring tool and should be part of any system administrator's toolset. You can monitor both VMware and your Backup infrastructures, which allows you to keep your systems running optimally and also detect problems that can be fixed. It is highly recommended that you try it out for yourself, while using this chapter as a starting point for configuration.

In this chapter, we're going to cover the following main topics:

- Exploring Veeam ONE – an overview
- Understanding Veeam ONE – installation and configuration
- Discovering monitoring – vSphere, vCloud, and Veeam
- Accessing reports
- Exploring Veeam ONE for troubleshooting

Let's get started!

Technical requirements

For this chapter, you should have Veeam Backup & Replication installed. *Chapter 1, Installation – Best Practices and Optimizations*, covered the installation and optimization of Veeam Backup & Replication, which you can leverage in this chapter. You will also need the ISO file for Veeam ONE to go through the installation process. You can obtain this from the Veeam website, in the downloads section: `https://www.veeam.com/virtualization-management-one-solution-download.html`.

Exploring Veeam ONE – an overview

Veeam ONE is part of the **Veeam Availability Suite** (a combination of Veeam Backup & Replication and Veeam ONE). It provides comprehensive monitoring and analytics for your backup, virtual, and physical environments. Veeam ONE uses Veeam Backup & Replication and Veeam Agents, plugins and backups for Nutanix AHV, and VMware vSphere and Microsoft Hyper-V to deliver intelligent monitoring, reporting, and automation for your environments. It uses interactive tools and intelligent learning to identify and resolve problems before they become a real issue in your environment.

Some of the critical capabilities of Veeam ONE include the following:

- **Built-in Intelligence**: Allows you to identify and resolve common infrastructure and software misconfiguration problems before they impact your operations
- **Governance & Compliance**: Allows you to monitor and report on SLA compliance for backups and data protection
- **Intelligent Automation**: Uses machine learning diagnostics and remediation actions to allow you to resolve issues faster before problems occur
- **Forecasting and planning**: Allows you to forecast both the costs and utilization of resources to determine future resource requirements

As well as these capabilities, Veeam ONE also offers many other features, including the following:

- **Proactive Alerting**: You can proactively mitigate potential threats before they become real problems
- **Chargeback and Billing**: Allows you to calculate both compute and storage costs per user/group for customer billing
- **Monitoring and Reporting**: Allows you to monitor backup, physical, and virtual environments 24/7, including email alerts and actions

- **Capacity Planning and Forecasting**: Allows you to forecast infrastructure requirements when planning future purchases

With Veeam ONE, many environments are supported within the application, such as the following:

- **Veeam Backup & Replication**: For monitoring your backup environment
- **VMware**: For monitoring vCenter, Hosts, vCloud, and Storage
- **Hyper-V**: For monitoring your hosts and storage
- **Nutanix AHV**: For monitoring AHV backups by Veeam
- **Microsoft Windows**: For monitoring Veeam Windows Agents
- **Linux**: For monitoring Veeam Linux Agents
- **AWS & Azure**: For monitoring cloud resources, backups, and Agents

See the *Further reading* section for a PDF link to everything new within Veeam ONE v10. The list is relatively extensive, and some of the big highlights are as follows:

- **NAS Backup**: Visibility into the File Servers and backups for NAS.
- **Nutanix AHV**: Monitoring backups for Nutanix AHV.
- **Cloud Visibility**: You can see all of your infrastructure and data centers from one console across all Cloud infrastructures, regardless of whether you're on-premises or cloud-based, for VMware, Hyper-V, AWS, and Azure.

Veeam ONE, when installed, is comprised of two main components that work together to provide monitoring and reporting capabilities:

- **Veeam ONE Monitor** is the primary tool used to monitor your virtual or physical environment and Veeam Backup & Replication infrastructure. Within the console, you can manage, view, and interact with alarms or monitoring data. You can also analyze the virtual and Backup infrastructure components' performance, including tracking, troubleshooting issues, generating reports, and administering monitor settings.
- **Veeam ONE Reporter** provides dashboards and reports for your entire environment. It allows you to verify configuration problems, optimize resources, track implemented changes, plan capacity growth, and ensure mission-critical workloads are protected.

These components get installed from the ISO file for Veeam ONE in a single solution. They can also be used in a distributed solution where components are separated, including the SQL Server requirements. The following diagram depicts the installation as one server solution (**Typical Deployment**) or via the distributed method (**Advanced Deployment**), with features residing on different systems:

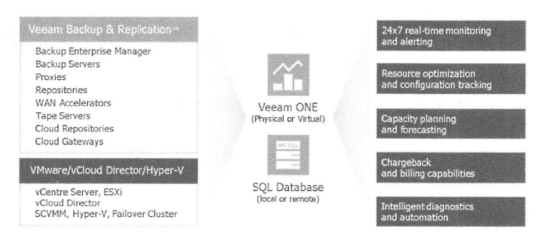

Figure 10.1 – Veeam ONE single server solution – typical deployment

The preceding diagram outlines the single server solution. We will outline some of the components and requirements for this in the next section.

Single server architecture – typical deployment

The following points outline some of the main options and components for a single server solution:

- Smaller environment – Veeam ONE server, Web UI, and Monitor client all installed on a single system.

- SQL Server express used on the same server, which saves licensing costs due to it being a smaller environment (SQL Server Express 2016 included).

- The virtual environment has less than 1,000 **virtual machines** (**VMs**) to monitor.

- Ability to install the Veeam ONE Monitor client on multiple machines for multi-user access.

There's also a distributed architecture where all the components are separated:

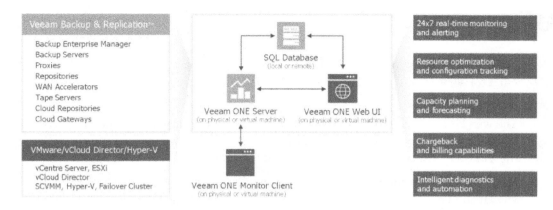

Figure 10.2 – Veeam ONE distributed architecture – advanced deployment

The preceding diagram outlines the distributed architecture solution. We will outline some of the components and requirements for this in the next section.

Distributed server architecture – advanced deployment

The following points outline some of the main options and components for a distributed architecture solution:

- Enterprise environment with servers distributed across multiple sites or data centers.

- Components are separated – the Veeam ONE Server and Web UI components are installed on separate servers.

- The database is typically a SQL Server Standard/Enterprise on a separate server – you can also enable **SQL Server Reporting Services** (**SSRS**) and leverage this within Veeam ONE to help with report creation.

- Typically, this is an environment that contains 1,000+ VMs that need to be monitored.

- You can install the Veeam ONE Monitor client on multiple systems for multi-user access.

> **Important note**
> For larger-scale deployments (1,000+ VMs), it is recommended that you use a remote SQL server installation. Veeam also recommends running the Veeam ONE services on a dedicated server. Using this type of distributed structure will improve the performance of Veeam ONE services.

It is also worth noting that the advanced installation relies on a client-server model for communication and data collection:

- The server components collect data from virtual infrastructure servers, vCloud Director servers, and Veeam Backup & Replication servers and stores this data in a SQL database.

- The web UI component (Veeam ONE Reporter) communicates with the SQL database to allow users to pull data for creating reports. It also communicates with the Veeam ONE server to determine what data is displayed based on its license type.

- Monitor client communicates with the Veeam ONE server to obtain real-time virtual and backup performance data and protection statistics.

> **Important note**
> To have a successful Veeam ONE deployment, you must ensure the client components are aware of the Veeam ONE server and the database's location so that it can connect and retrieve/process data.

Another thing to consider when you're deciding to deploy Veeam ONE is licensing. If you do not deploy a license during or after the installation, Veeam ONE operates in **Community Edition (Free)**. If you obtain a license, this is done using either **Per Socket Licensing** or **Per Instance Licensing**. An explanation of each license type is as follows. The *Further reading* section contains a link to the Veeam website in regards to licensing:

- **Per Socket Licensing**: This licensing type is based on the number of CPU sockets that are being managed by your VMware vSphere or Microsoft Hyper-V hosts. A license for each socket that the hypervisor can see is required.

- **Per Instance Licensing**: This licensing type is based on an *instance*, which is a unit (or token) that's assigned to a single object to make it manageable by Veeam ONE.

Furthermore, licensing is also different when monitoring Veeam Backup & Replication and Veeam Cloud Connect. Since they operate differently, there is also a license for each of these with Veeam ONE. Also, note that they cannot be installed on the same server, so if you need to monitor both Veeam Backup & Replication and Veeam Cloud Connect, you need to have two separate Veeam ONE servers. Keep this in mind when you're planning your deployment!

Lastly, there are a few other deployment planning and preparation things to consider regarding supported virtualization platforms, integration with vCloud Director, integration with Veeam Backup & Replication, limitations, and more. Please see the link entitled *Deployment Planning and Preparation* in the *Further reading* section at the end of this chapter for more details.

> **Important note**
> The following components *cannot* be installed on a Domain Controller: Veeam ONE Server, Veeam ONE Web UI, Veeam ONE Database, and Veeam ONE Agent. Also, any network configurations that use IPv6 addressing are not supported.

With that, we have covered what Veeam ONE is, what components make up its infrastructure, and other architectural information. Now, let's take a look at how to install and configure Veeam ONE.

Understanding Veeam ONE – installation and configuration

Installing Veeam ONE is a pretty straightforward process, and you are required to have the required ISO file, as noted in the *Technical requirements* section at the beginning of this chapter. Also, before installing, you need to check that you have the following prerequisites:

- **Check platform and system requirements**: Ensure that your virtual platform is supported and that the machine you are installing Veeam ONE on meets its hardware and software requirements. If you're using an advanced deployment with components on different servers, ensure they meet Veeam ONE's needs too.

- **Check account permissions**: Make sure the user account that Veeam ONE gets installed under has sufficient rights.

- **Check ports**: Make sure that all the required ports are open for communication between Veeam ONE's components, virtual infrastructure servers, vCloud Director servers, and Veeam Backup & Replication servers.

- **[Optional] Precreate Veeam ONE database**: Typically, the installation process for Veeam ONE will automatically create the SQL database for you; however, there could be times when it is necessary to create the database ahead of time. You can use the SQL script included within the Veeam ONE installation image for this.

Please see the link in the *Further reading* section for more information on *Deployment Planning and Preparation.*

The installation I will be using to demonstrate how to set up Veeam ONE will be the *typical deployment* due to my lab environment. As we discussed earlier, the *advanced deployment* is used when you're splitting the Veeam ONE components into different servers, and this depends on your environment and infrastructure.

For most businesses, scalable deployments of Veeam ONE would function as server applications. Smaller environments can run Veeam ONE on Windows 7 SP1, Windows 8, 8.1, and Windows 10.

Installing Veeam ONE v10

Before installing **Veeam ONE**, we need to ensure that we have a server deployed, either *Windows 2016 or 2019*, with enough disk space for the installation. The disk layout should be similar to the following:

- **OS drive**: This is where your operating system resides and should be used only for this purpose.

- **Application drive**: This will be your application installation drive for Veeam ONE and all its components.

- **Database drive**: Veeam ONE uses a SQL database to store its information, so installing the database files on a separate drive is advisable.

- **Performance cache drive**: Veeam ONE, when collecting performance data from the virtual infrastructure, stores it in a cache folder.

Once your server is ready and you have downloaded the **ISO file** and mounted it, perform the following steps to complete the installation:

1. Run the `setup.exe` file on the mounted ISO drive:

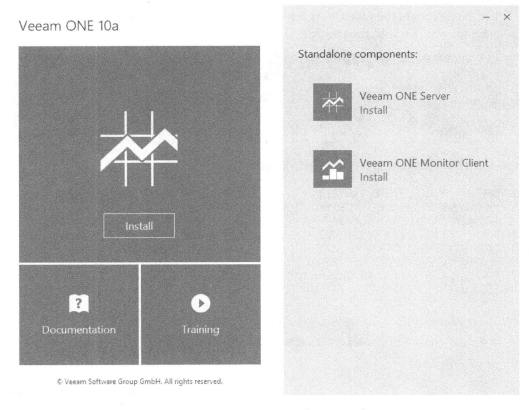

Figure 10.3 – Veeam ONE installation window

2. Click the **Install** button under the **Veeam ONE 10a** section on the left, or select the individual components that are used in an advanced deployment via **Install** under **Veeam ONE Server**.

3. The first screen is the **License Agreement** screen, so check both boxes and click **Next** to continue with the wizard:

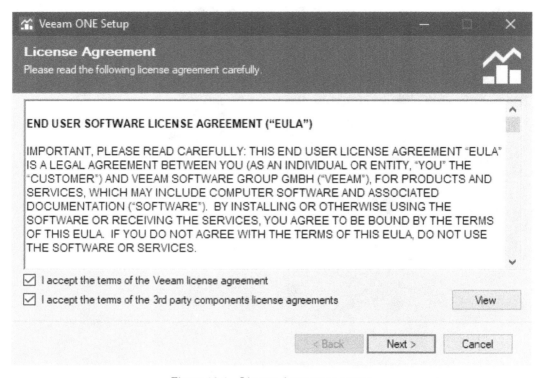

Figure 10.4 – License Agreement screen

4. The next screen is where you can select the deployment type, either **Typical** or **Advanced**. There's also a link at the bottom of the wizard that opens the Veeam ONE documentation website, which explains each in more detail:

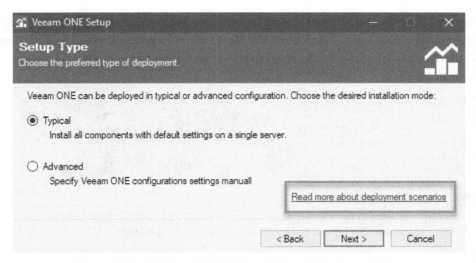

Figure 10.5 – Selecting the Typical or Advanced deployment type

5. Once you have selected your correct deployment type, click the **Next** button. We are using **Typical** for this particular installation due to lab limitations. You will now be taken to the **System Configuration Check** section of the wizard. Like the Veeam Backup & Replication installation, you can click the **Install** button to deploy any missing features:

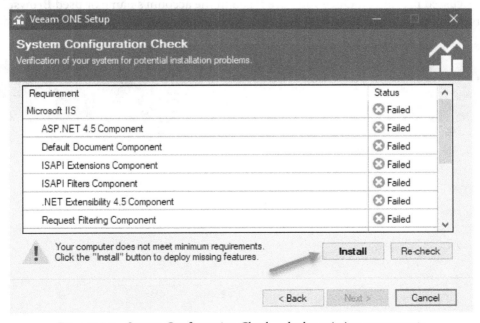

Figure 10.6 – System Configuration Check – deploy missing components

6. You can click the **Install** button to have the installation wizard deploy the missing components. Once completed, click the **Next** button to proceed to the **Installation Path**, which is where you will choose your server's application drive. Then, click **Next** to proceed:

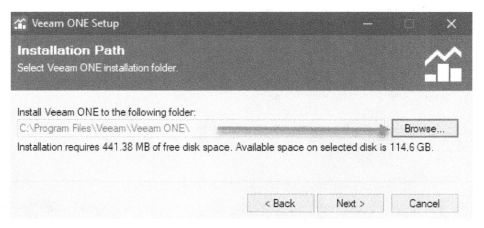

Figure 10.7 – Installation Path selection

7. On the next screen, you will be presented with the **Service Account Credentials** information you must use for Veeam ONE's installation. Account permissions were discussed previously. Once you have typed in the account's name or used **Browse**, type in the required **Password** and click **Next** to proceed:

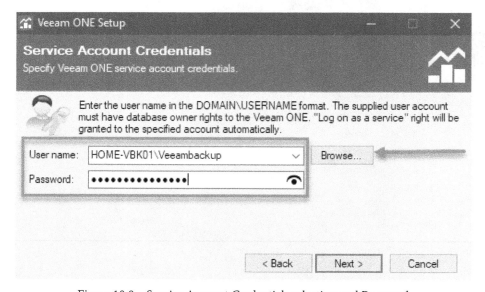

Figure 10.8 – Service Account Credentials selection and Password

8. The next screen is the **SQL Server Instance** section, and is where you can install either SQL Express locally or direct the installer to a remote SQL Server. The default database name is **VeeamONE**, but this can be changed. You can also select an authentication method for SQL, which will be either Windows or SQL Server authentication. Once you have set these options, click **Next** to proceed:

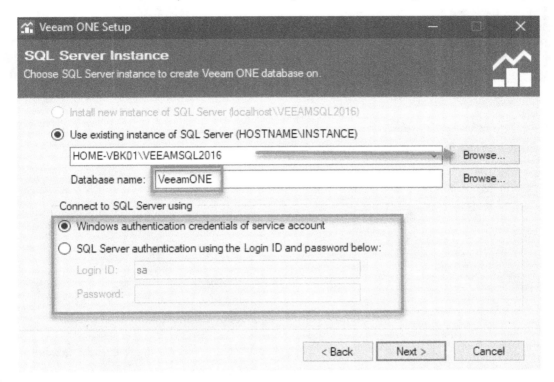

Figure 10.9 – SQL Server Instance selection

9. The next screen is the **Provide License** screen, and is where you can specify your license or select **Community Edition mode**. There is a link on this screen that specifies the limitations of the Community Edition. Once you have selected a license, click **Next** to proceed:

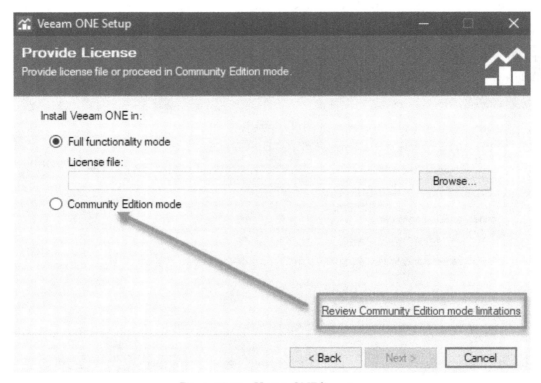

Figure 10.10 – Veeam ONE licensing

10. You will now be taken to the **Wizard's Connection Information**, which displays the default ports to use for **Reporter** and **Veeam ONE agent**. The certificate to be used is also noted here. This defaults to self-signed, but you can choose a different one from the drop-down list or by clicking the **View Certificate** button. Click the **Next** button to proceed:

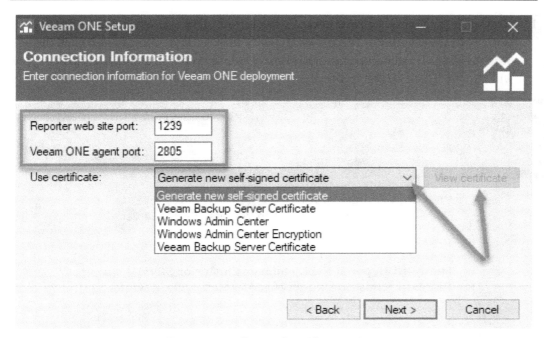

Figure 10.11 – Port and certificate settings

11. You will now be taken to the **Performance Data Caching** screen, which is where you can select the folder where you will store any performance data that's collected from the virtual infrastructure. Typically, as noted at the beginning of this section, that should be a separate drive on the server. Click **Next** once you have chosen a folder:

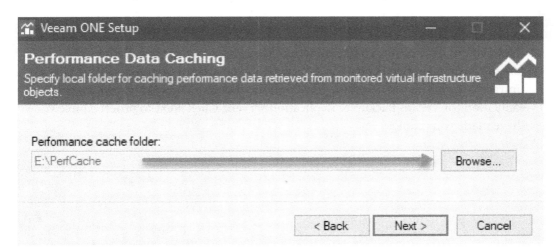

Figure 10.12 – Performance Data Caching directory

12. Next up is the **Virtual Infrastructure Type** section. This is where you can specify your **VMware vCenter** or **Microsoft Hyper-V** host settings. You can also click the **Skip virtual infrastructure configuration** option and complete this later, which we will do here. Click **Next** to proceed:

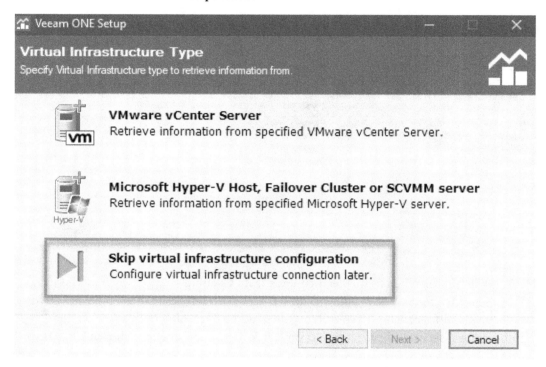

Figure 10.13 – Virtual Infrastructure Type selection during installation

13. Now, you must choose a **Data Collection Mode** based on the installation type you are performing, either **Typical** or **Advanced**. If you are not monitoring your virtual infrastructure, you can select the third option – **Backup Data Only** – which will only monitor Veeam Backup & Replication servers. Click **Next** to proceed once you have made your choice:

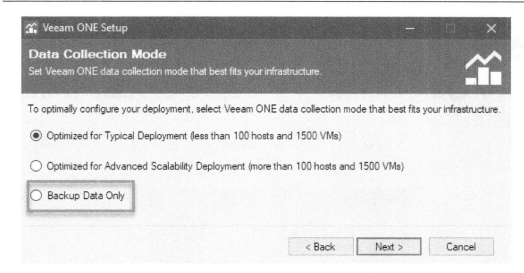

Figure 10.14 – Data Collection Mode selection

14. Now, the **Ready to Install** part of the wizard, along with the Veeam ONE configuration settings, will be displayed. A checkbox stating **Check for updates once the product is installed and periodically** is automatically selected here. This will check for updates occasionally. Click the **Install** button to begin installing Veeam ONE:

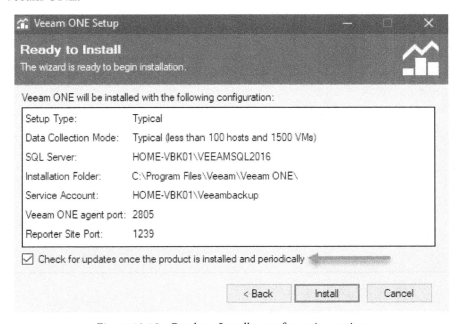

Figure 10.15 – Ready to Install – configuration review

With that, we have installed *Veeam ONE*. In the next section, we will learn how to configure monitoring for VMware and Veeam Backup & Replication.

Discovering monitoring – vSphere, vCloud, and Veeam

To launch the Veeam ONE programs you have installed, you can use the respective icons on your desktop:

Figure 10.16 – Veeam ONE icons

These icons launch the respective programs that were installed alongside Veeam ONE:

- **Veeam ONE Monitor**: This launches the monitor console, which is where you configure and monitor your virtual infrastructure, vCloud Director, and Veeam Backup & Replication, and configure alarms.

- **Veeam ONE Reporter**: This launches your web browser and takes you to your Dashboards, Workspace, and Configurations.

When you first launch the Veeam ONE Monitor client, a small wizard will appear, asking for both your **SMTP (Email)** and **SNMP settings**. The following screenshot shows the initial screen and the different tabs you must fill out to complete this wizard. You can click **Cancel** to skip this. You can configure the email and SNMP settings later in the console:

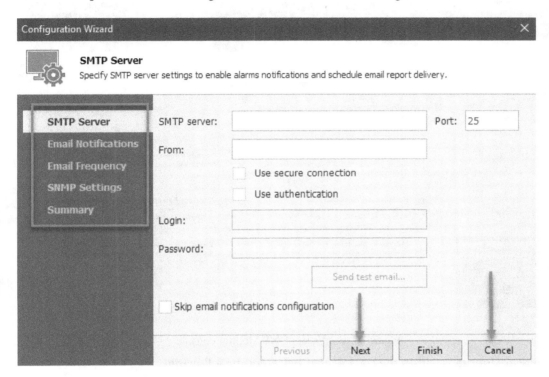

Figure 10.17 – Configuration Wizard on the first launch of the program – SMTP and SNMP Settings

When the **Veeam ONE Monitor** client opens, you will be presented with the central console view of Veeam ONE. Here, you can launch the Notification wizard shown in the preceding screenshot using the **Notifications** button shown in the following screenshot:

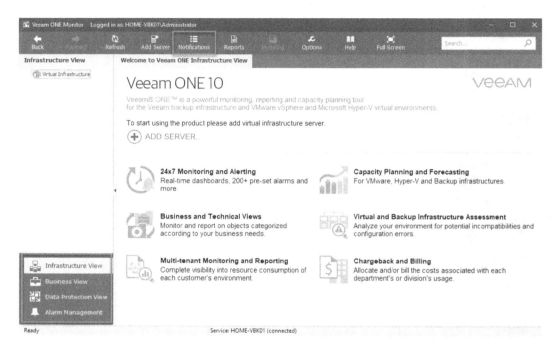

Figure 10.18 – Veeam ONE Monitor console

There are four tabs within the console that represent the following:

- **Infrastructure View**: This shows either your VMware, vCloud Director, or Hyper-V environment once you've added the respective servers to Veeam ONE.

- **Business View**: This shows your virtual infrastructure from a business view perspective, based on your company's needs and priorities. You can group your virtual infrastructure objects by things such as business unit, department, purpose, SLA, and so forth.

- **Data Protection View**: This will show the view for the backup servers that get connected to the console; that is, either Veeam Backup & Replication or Veeam Cloud Connect.

- **Alarm Management**: This tab allows you to manage all the default alarms that come with Veeam ONE. You can also add custom alarms here.

To get started with monitoring your environment, follow these steps:

1. Click on the **+ ADD SERVER** option on the screen or the **Add Server** button in the
    toolbar. You can add any server from any screen as they are not related when you're
    doing this, but rather when you're looking at each tab:

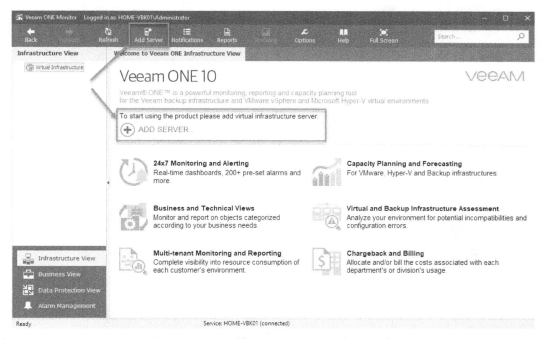

Figure 10.19 – Add Server options in the console

2. Once you click on either option, you will be presented with the **Add Server Wizard** screen, where you can add **VMWARE SERVER, VMWARE VCLOUD DIRECTOR, HYPER-V SERVER,** or **VEEAM BACKUP AND REPLICATION SERVER:**

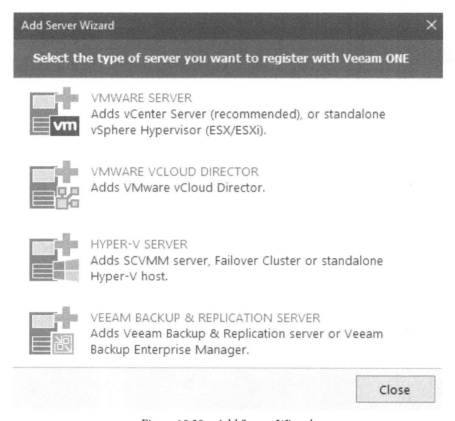

Figure 10.20 – Add Server Wizard

As an example, we will add both a VMware Server and Veeam Backup & Replication Server to see the process of setting them up in Veeam ONE.

3. First, click on the **VMware Server** option in the **Add Server Wizard** screen, as shown in the preceding screenshot. You will be presented with the **Add Server Wizard** screen:

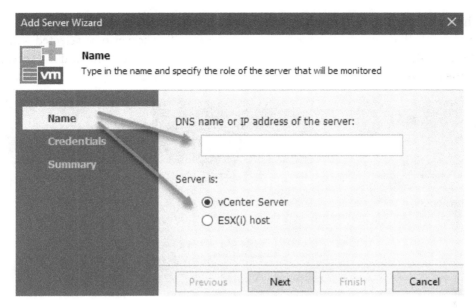

Figure 10.21 – Add Server Wizard – VMware Server

4. From this screen, enter the server's **DNS or IP address** and select whether it is **vCenter Server** or **ESX(i) host**. Click **Next** to proceed.

5. You will then get prompted for the **Credentials** information for either vCenter Server or the ESXi host. Enter the required credentials and click **Next** to continue:

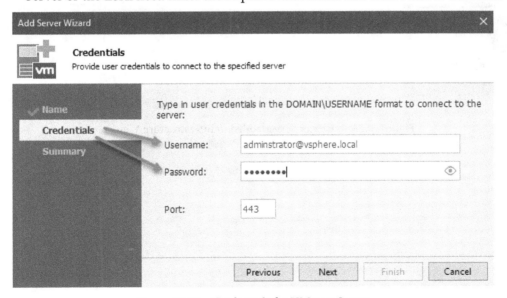

Figure 10.22 – Credentials for VMware Server

6. The final screen is the **Summary** screen, which shows the server's name and the credentials that must be used to connect to the specified server type selected – in this case VMware vCenter. Click **Finish** to complete the wizard and add the server to the console screen under the **Infrastructure View** section:

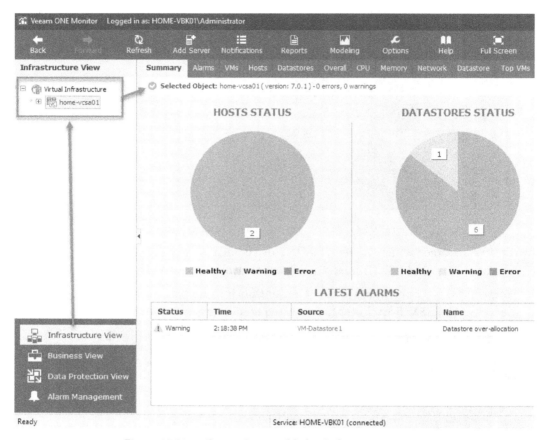

Figure 10.23 – vCenter Server added to Infrastructure View

Now, follow the same **Add Server Wizard** options but select **VEEAM BACKUP AND REPLICATION SERVER**. Enter the server name or IP address, add the necessary credentials, and complete the wizard to add your backup server to the console's **Data Protection View**:

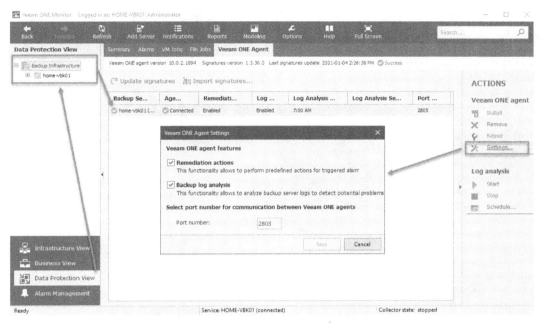

Figure 10.24 – Veeam Backup & Replication server added to Data Protection View

This screen shows the **Veeam ONE Agent** tab so that you can confirm the server's installation was completed correctly. Once it's been added and connected, you will see the **Summary** screen, which shows details including job status, repositories, and file and agent backups once data starts to be collected:

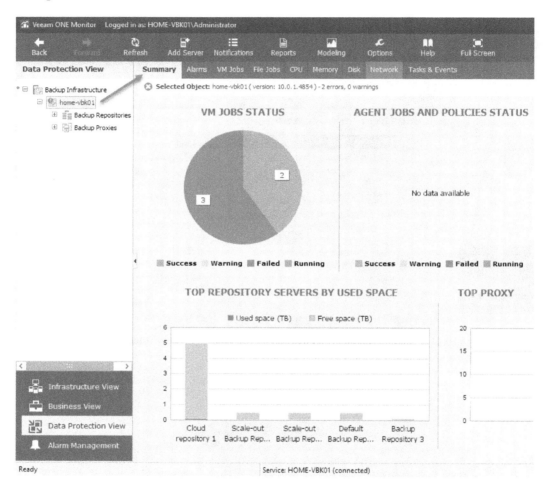

Figure 10.25 – Veeam Backup & Replication Server Summary

In this section, you learned how to add both your virtual servers and Veeam Backup & Replication servers to Veeam ONE. In the next section, we'll look at reporting using both the Veeam ONE Monitor client and Veeam ONE Reporter.

Accessing reports

One of the advantages of using Veeam ONE is the reporting ability that it provides. You can get reports on all types of environments that you add, both for virtual infrastructure and backups. Veeam ONE comes with many canned reports out of the box that suit most requirements. The reports will also change depending on whether you add VMware, Hyper-V, or Veeam Backup & Replication to your environment.

Some of the predefined report groups that are available are as follows:

- Nutanix AHV Protection
- Infrastructure Chargeback
- Veeam Cloud Connect
- Veeam Backup Assessment
- Veeam Backup Billing
- Veeam Backup Capacity Planning
- Veeam Backup Monitoring
- Veeam Backup Overview
- Veeam Backup Tape Reports
- Veeam Backup Agents
- Public Cloud Data Protection
- VMware Infrastructure Assessment
- VMware Overview
- VMware Monitoring
- VMware Optimization
- VMware Capacity Planning
- VMware Configuration Tracking
- Hyper-V Infrastructure Assessment
- Hyper-V Overview
- Hyper-V Monitoring
- Hyper-V Optimization
- Hyper-V Capacity Planning
- Custom Reports
- vCloud Director
- Offline Reports

Figure 10.26 – Default report groups for Veeam ONE

There are two ways to access reports within the Veeam ONE environment:

- **Veeam ONE Monitor**: You can open the monitoring console and access reports from there by highlighting something within the **Infrastructure View** and **Data Protection View** areas of the console and selecting the **Reports** button in the toolbar.

    Infrastructure View:

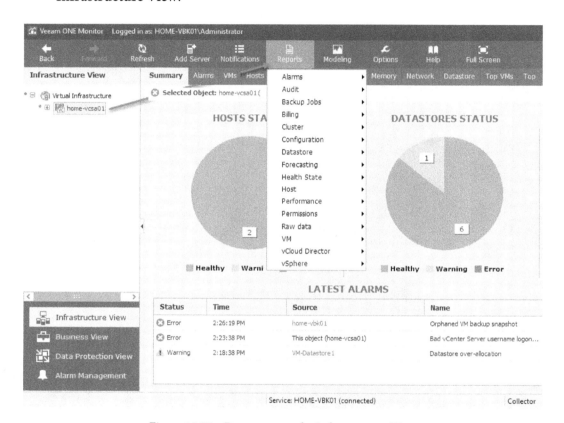

Figure 10.27 – Reports menu for Infrastructure View

Data Protection View:

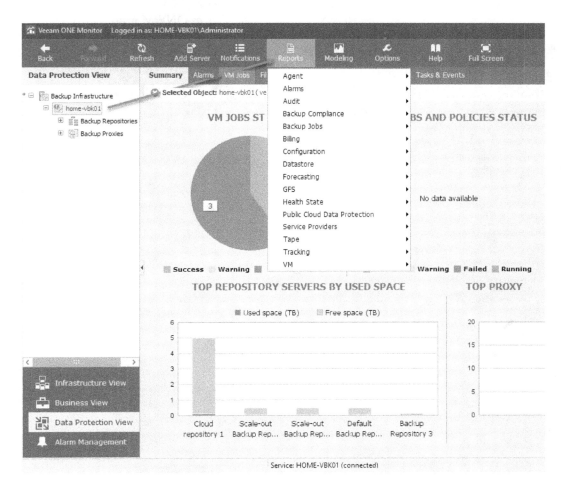

Figure 10.28 – Reports menu for Data Protection View

> **Important note**
> Depending on which view you are in, the **Report** menu options change regarding virtual infrastructure or backups.

- The other way to access reporting is by using the **Veeam ONE Reporter** icon, which launches a web browser interface. In the **WORKSPACE** tab, you can find a hierarchy of reports, similar to those found in the Veeam ONE Monitor console:

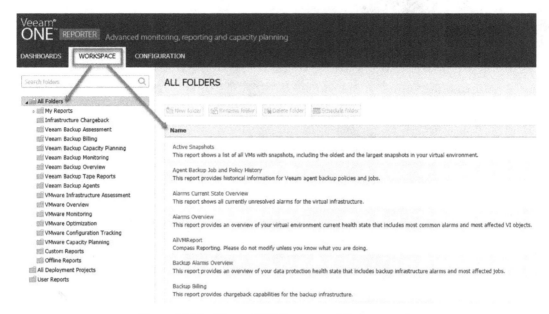

Figure 10.29 – Veeam ONE Reporter – Workspace tab

To view details of any report, click on the report's name, enter the required parameters, and then select either **Preview** or **Save As** to create your custom report with the already entered parameters. Reports get saved in the **My Reports** folder, and you can create subfolders to organize reports. The following is an example of the Backup Infrastructure Audit report:

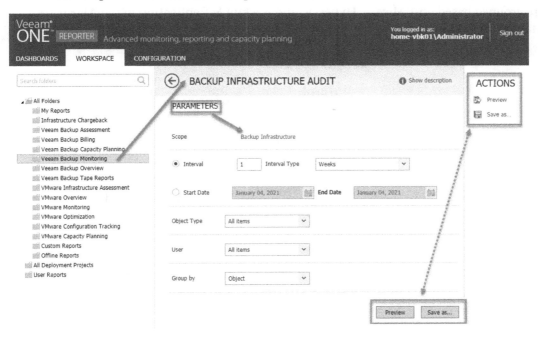

Figure 10.30 – Backup Infrastructure Audit report example

With that, we have covered the basic reporting capabilities of Veeam ONE. For our last topic, we'll look at how Veeam ONE can help with your environment's troubleshooting issues.

Exploring Veeam ONE for troubleshooting

When it comes to troubleshooting your environment, whether it be the virtual infrastructure or backups, this is where Veeam ONE shines. Veeam ONE uses **Signatures**, which are related to specific issues in their **Knowledge Base**, along with **Alarms** to alert you to potential problems, referred to as **Veeam Intelligent Diagnostics**. Alarms are grouped based on **VMware**, **Hyper-V**, and **Backup & Replication** in the **Alarm Management** tab:

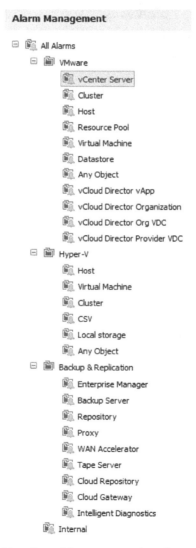

Figure 10.31 – Alarm Management tab – Alarm groupings

Alarms are triggered based on events and can alert you in various ways, such as via email (SMTP), SNMP, or running a script. You can also turn on automation so that Veeam ONE can fix the issue for you based on the rules you create. An example of this can be found under the **VMware | Virtual Machine** section for alarms, where there is an alarm called **VM with no backup**. This alarm checks all virtual machines in your infrastructure, and if it finds that the backup has not run within, say, 24 hours, it can automatically add it to a backup job:

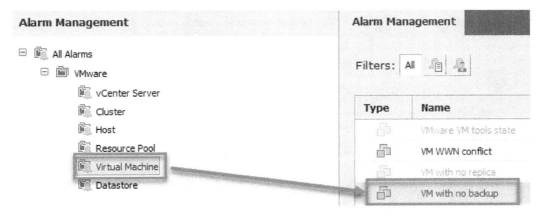

Figure 10.32 – VM with no backup alarm

The following are the properties of an alarm that automates the backup:

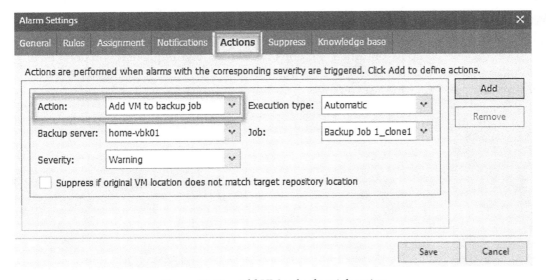

Figure 10.33 – Add VM to backup job action

Along with adding to a backup job, there are multiple other **Actions** options:

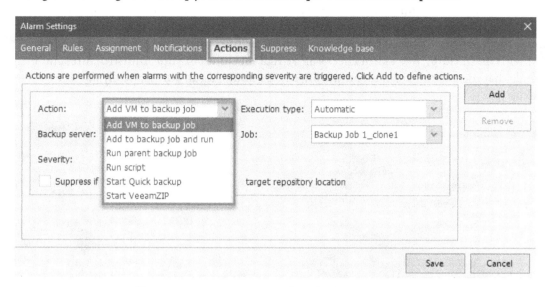

Figure 10.34 – Action options for a VM with no backup

Now, let's look at how we can see these options right from within the Veeam ONE Monitor console. If you go to the **Infrastructure View** tab and highlight your server, you will see the **VM with no backup** alarm present on the right-hand side of your lab environment.

What this alarm means is that your VM has no backup based on some checks that Veeam ONE has completed. If, using the rules in Veeam ONE, you have automated any VMs that are not being backed up to jobs, then you don't need to do anything, but if you don't have automation set up, then it is recommended that you add this VM to a backup job on your Veeam Backup & Replication server:

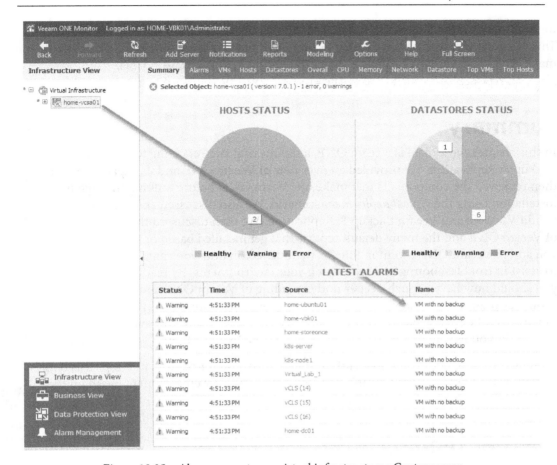

Figure 10.35 – Alarm present on a virtual infrastructure vCenter server

With that, we have covered the last part of this chapter regarding how to use Veeam ONE to help troubleshoot and remediate your environment.

Veeam ONE community resources

The Veeam Product Strategy Team has some dynamic resources available so that you can do more with Veeam Backup & Replication, Veeam ONE, and other Veeam products. Be sure to follow #DRTestTuesday on Twitter to learn more about how Veeam Availability Orchestrator simplifies your DR testing, planning, and documentation:

```
https://twitter.com/search?q=%23DRTestTuesday&src=typeahead_
click&f=live
```

Be sure to follow the Veeam ONE Catch of the Day on Twitter with #VeeamONECOTD. This is a great way to learn how to use Veeam ONE practically, as well as make other discoveries regarding Veeam ONE: https://twitter.com/search?q=%23VeeamONECOTD&src=typed_query&f=live.

Summary

In this chapter, we looked at Veeam ONE, a monitoring and reporting tool for Veeam Backup & Replication. We provided an overview of Veeam ONE and its capabilities and then reviewed the components that make up Veeam ONE, before walking through its installation using the *typical deployment* scenario. We also discussed and showed you how to add VMware and Veeam Backup & Replication. We then discussed the reporting aspect of Veeam ONE and the many default reports that get installed based on what servers you add to Veeam ONE Monitor. Finally, we touched on how you can use Veeam ONE to assist in troubleshooting and remedying your environments. By reading this chapter, you should now have a much deeper understanding of Veeam ONE and what it is. You should also understand how to install and set up servers for monitoring and reporting. Finally, you should be able to utilize Veeam ONE to assist with your environment's triage. Hopefully, you now have a better idea of how Veeam ONE can fit into your environment.

This chapter concludes this book, and I want to thank you for reading it. I hope this book's content will be beneficial to you and help you configure your environment based on the topics that were covered.

Further reading

- Deployment Scenarios: https://helpcenter.veeam.com/docs/one/deployment/deployment_scenarios.html?ver=100

- Veeam ONE Licensing: https://helpcenter.veeam.com/docs/one/deployment/licensing.html?ver=100

- Deployment Planning and Preparation: https://helpcenter.veeam.com/docs/one/deployment/deployment_planning_preparation.html?ver=100

- Veeam ONE Monitor User Guide: https://helpcenter.veeam.com/docs/one/monitor/about.html?ver=100

- Veeam ONE Reporter User Guide: https://helpcenter.veeam.com/docs/one/reporter/about.html?ver=100

- Working with Alarms: `https://helpcenter.veeam.com/docs/one/alarms/about.html?ver=100`

- Multi-Tenant Monitoring and Reporting: `https://helpcenter.veeam.com/docs/one/multitenant/about.html?ver=100`

- Veeam ONE – Default Reports: `https://helpcenter.veeam.com/docs/one/reporter/predefined_reports.html?ver=100`

- Veeam ONE v10 – What's New: `https://www.veeam.com/veeam_one_10_0_whats_new_wn.pdf`

Packt.com

Subscribe to our online digital library for full access to over 7,000 books and videos, as well as industry leading tools to help you plan your personal development and advance your career. For more information, please visit our website.

Why subscribe?

- Spend less time learning and more time coding with practical eBooks and Videos from over 4,000 industry professionals

- Improve your learning with Skill Plans built especially for you

- Get a free eBook or video every month

- Fully searchable for easy access to vital information

- Copy and paste, print, and bookmark content

Did you know that Packt offers eBook versions of every book published, with PDF and ePub files available? You can upgrade to the eBook version at packt.com and as a print book customer, you are entitled to a discount on the eBook copy. Get in touch with us at customercare@packtpub.com for more details.

At www.packt.com, you can also read a collection of free technical articles, sign up for a range of free newsletters, and receive exclusive discounts and offers on Packt books and eBooks.

Other Books You May Enjoy

If you enjoyed this book, you may be interested in these other books by Packt:

Mastering Azure Security

Mustafa Toroman, Tom Janetscheck

ISBN: 978-1-83921-899-6

- Understand cloud security concepts
- Get to grips with managing cloud identities
- Adopt the Azure security cloud infrastructure
- Grasp Azure network security concepts
- Discover how to keep cloud resources secure
- Implement cloud governance with security policies and rules

AWS Security Cookbook

Heartin Kanikathottu

ISBN: 978-1-83882-625-3

- Create and manage users, groups, roles, and policies across accounts
- Use AWS Managed Services for logging, monitoring, and auditing
- Check compliance with AWS Managed Services that use machine learning
- Provide security and availability for EC2 instances and applications
- Secure data using symmetric and asymmetric encryption
- Manage user pools and identity pools with federated login

Packt is searching for authors like you

If you're interested in becoming an author for Packt, please visit `authors.packtpub.com` and apply today. We have worked with thousands of developers and tech professionals, just like you, to help them share their insight with the global tech community. You can make a general application, apply for a specific hot topic that we are recruiting an author for, or submit your own idea.

Leave a review - let other readers know what you think

Please share your thoughts on this book with others by leaving a review on the site that you bought it from. If you purchased the book from Amazon, please leave us an honest review on this book's Amazon page. This is vital so that other potential readers can see and use your unbiased opinion to make purchasing decisions, we can understand what our customers think about our products, and our authors can see your feedback on the title that they have worked with Packt to create. It will only take a few minutes of your time, but is valuable to other potential customers, our authors, and Packt. Thank you!

Index

S

www.ingramcontent.com/pod-product-compliance
Lightning Source LLC
Chambersburg PA
CBHW062101050326
40690CB00016B/3168